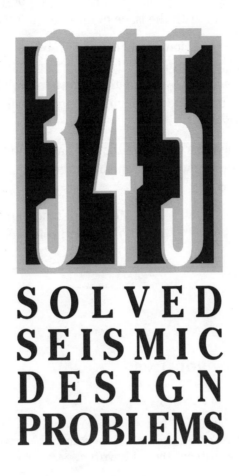

# SOLVED SEISMIC DESIGN PROBLEMS

**Fourth Edition**

**Majid Baradar, PE**

Professional Publications, Inc.
Belmont, CA

# How to Get Online Updates for This Book

I wish I could claim that this book is 100% perfect, but 25 years of publishing have taught me that engineering textbooks seldom are. Even if you only took one engineering course in college, you are familiar with the issue of mistakes in textbooks.

I am inviting you to log on to Professional Publications' web site at **www.ppi2pass.com** to obtain a current listing of known errata in this book. From the web site home page, click on "Errata." Every significant known update to this book will be listed as fast as we can say "HTML." Suggestions from readers (such as yourself) will be added as they are received, so check in regularly.

PPI and I have gone to great lengths to ensure that we have brought you a high-quality book. Now, we want to provide you with high-quality after-publication support. Please visit us at **www.ppi2pass.com**.

Michael R. Lindeburg, PE
Publisher, Professional Publications, Inc.

**345 SOLVED SEISMIC DESIGN PROBLEMS**
**Fourth Edition**

Copyright © 2000 by Professional Publications, Inc. All rights reserved. No part of this publication may be reproduced, stored in a retrieval system, or transmitted, in any form or by any means, electronic, mechanical, photocopying, recording, or otherwise, without the prior written permission of the publisher.

Printed in the United States of America

Professional Publications, Inc.
1250 Fifth Avenue, Belmont, CA 94002
(650) 593-9119
www.ppi2pass.com

Current printing of this edition:   3

**Library of Congress Cataloging-in-Publication Data**
Baradar, Majid, 1953-
    345 solved seismic design problems / Majid Baradar. -- 4th ed.
        p.    cm.
    Includes bibliographical references (p.    ).
    ISBN 1-888577-35-5
    1. Earthquake resistant design--Problems exercises, etc.
  2. Earthquake resistant design--Examinations Study guides.
  I. Title.  II. Title: Three hundred forty-five solved seismic design
  problems.
  TA658.44.B37   1999
  624.1′762--dc21                                      99-14452
                                             CIP

The passion for excellence begins today
and determines tomorrow's achievement.

Dedicated to my father who enlightened
others with his accomplishments and
excellence, championed others with his
trust and vision, and helped others with
his superior commitment and devotion.

Majid Baradar

# TABLE OF CONTENTS

# PREFACE TO THE FOURTH EDITION

Since early 1995, the extensive use of this book in college classes and technical schools, and by practicing civil engineers, PE exam candidates, architects, and other professionals, has encouraged me to continue (in this fourth edition) with the original intent of the book—to introduce exam-type problems to supplement seismic design text books and the fundamental earthquake engineering of structures.

Each new edition of this book reflects the latest changes in the *Uniform Building Code* (UBC), trends and practices in seismic engineering, and the newest developments in the civil engineering seismic examination. Recently, the 1997 UBC was published. For the development of a national code by the year 2000, the 1997 NEHRP Provisions serve as the source document for the 1997 UBC and other U.S. model building codes (BOCA and SBC). The revised edition of this code incorporates changes approved since the last edition (1994). The seismic provisions in the 1997 UBC have been extensively revised from the previous (1994) edition. For that reason, this fourth edition of *345 Solved Seismic Design Problems* has been updated to conform to the 1997 UBC provisions.

As noted in previous editions, my motivation to begin this book was the lack of any detailed test plans identifying what knowledge was necessary for competent entry-level civil engineering practice in the area of seismic principles. *345 Solved Seismic Design Problems* strives to serve that purpose. In 1995, it was very interesting and inspiring for me to observe the examination test plan developing and changing. In 1996, the Board of Registration launched a well-organized test plan for the Special Civil Engineer Seismic Examination. To the fullest extent possible, the test plan identifies the necessary tasks and knowledge related to seismic principles.

I am delighted with the test plan, as well as overwhelmed by the task of making sure that my book adequately covers all of the seismic principles areas for entry-level civil engineers. *345 Solved Seismic Design Problems* is entirely consistent with the current test-plan-defined content areas. The organization and purpose of this edition are similar to the first edition, but the fourth edition conforms to the provisions of the 1997 UBC.

In addition, a great effort was made to offer this edition in both English and SI units; it has been rewritten in its entirety to provide you a choice of units. I am proud to serve the needs of civil engineers with my work.

This new edition is offered to the user in the same spirit as were the earlier editions—to make a contribution to the goals of public safety in the event of an earthquake. Words cannot express my heartfelt gratitude to all those who so generously and professionally offered their knowledge and support to me over the last few years. Be certain that your success in the examination and in your career is always my wish.

Good Luck!

Majid Baradar
Pinole, CA

# ACKNOWLEDGMENTS

I would like to express, first and foremost, my appreciation to all the engineers, architects, other professionals, and friends who offered their experiences, feedback, joy, and acknowledgments for this publication.

Next, special thanks go to Ed Haverlah, SE, who reviewed and recommended my original book to PPI for publishing. I am also grateful to Michael R. Lindeburg, PE—PPI's president—for his approval of this book throughout its preparation.

To acquisitions editor Jason Standifer, who first envisioned the publication of this book; to the retired acquisitions editor Gerald R. Galbo, for his confidence in me from the very beginning; to acquisitions editor Tom Tolfa, for his invaluable assistance with the fourth edition; to copy editors Jessica R. Whitney-Holden and Vanessa R. Franco, for their constant support; to Sylvia M. Osias and Kate Hayes, for their attentiveness at typesetting; to Mia Laurence, Jessica R. Whitney-Holden, and Aline S. Magee, for their accuracy in proofreading; and to Gregory J. McGreevy, for his work on creative design, many thanks.

I want to profess my deepest appreciation to my mother, F. Shahifakhr, who, through her love and confidence, has empowered me. In preparing this book for publication, I have received encouragement and support from other family and friends, as well. Notable among them are my sister, Nahid Baradar, for believing in me; my brothers, for supporting and understanding the value of this publication to me; and my son, Salar, who was most proud to see his father's name on this book—I am very pleased to be his father.

A very special thanks is extended to my wife, Mitra, for her patience, love, and understanding over the many years it took to make this book a reality.

Most of all, my immeasurable appreciation to my father, who loved learning and teaching and who was my inspiration.

# INTRODUCTION

This section introduces you to the seismic exam and tells you how to use *345 Solved Seismic Design Problems*. It also explains how to relate this book to reference materials, such as the *Uniform Building Code, Seismic Design of Building Structures,* and others. Using this book as a study guide, you will be surprised to discover how effectively you can learn seismic principles and fundamental earthquake engineering of structures to successfully pass the special seismic exam.

# THE SEISMIC EXAM

## BRIEF BACKGROUND

In April 1988, the State Legislature mandated that the California Board for Professional Engineers and Land Surveyors test all applicants/civil engineers in the seismic principles area. To accomplish this mandate, the Board set forth Special Civil Engineer Seismic Principles Examination standards to reliably test professional registration applicants' competency to provide safe engineering services. The special seismic principles section of the Professional Engineer (Civil) licensing examination responds to the public's increasing awareness that structures in seismically active areas need to be designed to resist seismic forces. The mission of the Board is to safeguard the public from incompetent practice. To become registered, the PE candidate must obtain a passing score on this special exam.

## EXAM CONTENTS

To test a candidate's knowledge of seismic principles and fundamental earthquake engineering, the California Board for Professional Engineers, in its June 15, 1990, meeting, adopted the original test plan areas for the special seismic exam. The Board distributes the plan in an effort to better inform candidates on important examination subject areas. Recently, the latest test plan was introduced in the Special Civil Engineer Examination Information for Examinees brochure. The test plan lists the examination content in terms of entry-level civil engineering tasks and knowledge statements.

## EXAM FORMAT

The exam is composed of about 45–50 multiple-choice questions drawn from subjects outlined in the Board-adopted test plan. All the questions should be answered. For each question, there may be more than one correct answer. In such cases, you should select

select the best correct answer. Mark only one answer to each question. Record your answers in the official, computer-readable solution booklet. Credit will not be given for questions worked out or answered only in the official test booklet (as opposed to the solution booklet).

You will be given 2 1/2 hours to complete this exam. The total points will not be evenly distributed among the questions; they will vary from one question to another depending on the significance, difficulty, and complexity of each question. The official test booklet will state the point values assigned to each question.

## TIME MANAGEMENT DURING THE EXAM

The exam is a performance test. To be successful, in addition to demonstrating your understanding of seismic principles and fundamental earthquake engineering, you must manage your time well during the testing period. This will require you to be more focused during the exam than you were during your study periods.

As mentioned, the exam consists of about 45 questions that must be answered in 2 1/2 hours, or 150 minutes. A quick calculation based on these numbers shows that you have, on the average, 3 minutes and 20 seconds per question.

To make efficient use of your time:

1. Answer questions that can be answered immediately. Some questions may require minimal calculations.

2. Answer questions that you feel confident about but that may require more time for calculations.

3. Answer questions about which you have doubts but for which you know where to find relevant information in your books, references, or notes.

4. Answer questions that you are not sure of but believe you have a chance of getting right.

5. Answer all remaining questions by guessing. You are not penalized for wrong guesses, so don't leave any questions unanswered.

If you complete the exam with time to spare, recheck your calculations and answers. Verify calculations that involve unit conversions. Confirm any specific doubts by looking up relevant information in your books, references, and notes.

## EXAM GUIDELINES

The examination is open book. You may bring any number of books of any type into the exam. Loose-leaf notepaper, tables, and reference materials must be gathered into a binder. You might find it helpful, too, to bring *345 Solved Seismic Design Problems* to the exam along with any notes you have taken during your exam preparation.

It is essential that you bring a calculator to the exam. There are no restrictions on the type of calculator, although it should be battery-powered and silent and may not have a QWERTY keyboard. You will not be permitted to plug your calculator into an outlet during the examination.

You will not be permitted to share books, reference materials, calculators, or any other items with other candidates during the exam period. Speaking with other candidates is also prohibited. If a proctor suspects a candidate of not following the instructions, the candidate may be expelled from the examination. Disqualification from the exam could jeopardize your eligibility for future registration.

## EXAM DATES AND LOCATIONS

The seismic principles section of the Special Civil/Seismic Principle Examination is given twice a year, in mid-April and late October. The examination is scheduled for a Saturday and does not conflict with the April PE Civil Examination. The State Board will notify candidates of the date, time, and location of the examination. The Board may be contacted for more information.

Board for Professional Engineers and Land Surveyors
www.dca.ca.gov/pels

| | |
|---|---|
| Physical address: | 2535 Capitol Oaks Drive, Suite 300<br>Sacramento, CA 95833-2944 |
| Mailing address: | P.O. Box 349002<br>Sacramento, CA 95834-9002 |
| Telephone: | (916) 263-2222 |
| Fax: | (916) 263-2246 |
| E-mail: | bpels@dca.ca.gov |

# HOW TO USE THIS BOOK

## FORMAT AND CONTENT OF THIS BOOK

*345 Solved Seismic Design Problems* promotes and reinforces a candidate's knowledge of seismic principles and fundamental earthquake engineering of structures, the *Uniform Building Code* (UBC), and the seismic exam format and content. It is written as a study guide that mimics the seismic exam format. The questions reflect the examination subject areas and intent as stated by the Board. The exam provides a basic review of seismic principles and fundamental earthquake engineering of structures, with a focus on buildings.

How does *345 Solved Seismic Design Problems* help you?

- It gives you information that will expand your knowledge.
- It introduces you to the organization and "feel" of the exam.
- It gives you practice in answering questions that are presented in the same format as the exam.
- It provides insight into the approaches that may be used to solve exam problems.
- It acquaints you with the UBC.

*345 Solved Seismic Design Problems* presents all topic areas of the Board-adopted test plan in accordance with the level of understanding of seismic design principles that the Board has deemed necessary. Topics are organized into chapters composed of multiple-choice problems. Each chapter follows a programmed learning approach that begins with simple problems and then gets progressively more involved and complicated. Each sample question is fashioned to emphasize a key point of a specific topic area.

The problems are designed to convey the basic seismic engineering fundamentals, yet some require more knowledge, judgment, and insight than others. These problems are considerably more tedious and longer than those on the seismic exam, but they do cover necessary subject areas. All sample questions and problems have been developed to be consistent with UBC requirements.

*345 Solved Seismic Design Problems* provides relatively complete solutions to all practice questions and problems. For questions, solutions are limited to an explanation of the pertinent key points(s) and appropriate UBC reference(s). For problems, the essential steps to every solution are offered. Where applicable, SI units are presented along with their customary U.S. counterparts. The metric conversions are provided in parentheses following the English units, and they conform to current industry standards.

## TECHNIQUES AND STEPS FOR USING THIS BOOK

*345 Solved Seismic Design Problems* works with seismic textbooks and the *Uniform Building Code* (UBC). (See the following section, "How to Use this Book with Other References.") Follow the steps listed below to begin your exam preparation.

*Step 1*: Glance through each chapter. Familiarize yourself with the format of the sample questions. Do not attempt to answer any questions at this time.

*Step 2*: For each chapter, study the corresponding subjects from your textbook(s) and the UBC. You should visualize possible exam questions as you study. Do not be discouraged if you find it difficult to interpret the UBC provisions during this step.

*Step 3*: In each chapter, you will find a summary of the chapter's topics along with sample questions similar to those on the actual exam. Answer the sample questions, and compare your answers with the solutions provided. For recording your answers, you may use provided answer sheets for each chapter in this book.

If you find a subject that you are not clear about or acquainted with, refer to your text(s) and the UBC to examine that subject in depth.

*Step 4*: Review these chapters as many times as needed until you've thoroughly learned all their topics areas and questions.

The value of this book to you will depend on how many of its insights and techniques you can adopt as routine ways of thinking and applying knowledge. You'd be wise to review each chapter several times. The topics and the format of the exam should become second nature to you. Each review will help you identify additional areas that require your attention, so that by the day of the exam you will be fully prepared and confident.

## HOW TO USE THIS BOOK WITH OTHER REFERENCES

Appendix B of *345 Solved Seismic Design Problems* lists Board-recommended study references. Although the author suggests that you use *Seismic Design of Building Structures*, by Michael R. Lindeburg, PE, as your main resource, this book is devised to work with all seismic textbooks. Also, cross-references to the *Uniform Building Code* (UBC) are used throughout the book to direct you to more detailed information.

Provisions of the 1997 edition of the UBC and the *UBC Standards* have been separated and grouped into a three-volume set.

Volume 1—Administrative, Fire and Life-Safety, and Field Inspection Provisions
Volume 2—Structural Engineering Design Provisions
Volume 3—Material, Testing, and Installation Standards

Design standards are included in Volume 2 that previously were published in the UBC Standards. Design standards have been added to their respective chapters as divisions of the chapters. The UBC was metricated in the 1994 edition. The metric conversions are provided in parenthesis following the English units.

Note that the UBC is an interpretive document. While looking into the UBC provisions, you should investigate them thoroughly and inspect all possible exceptions and related criteria. Important UBC provisions and relevant exceptions are presented in this book to clarify their interpretation.

When applicable, *345 Solved Seismic Design Problems* provides UBC references for each sample problem. Each reference to the UBC consists of a chapter and section designation. While reviewing a particular section of this book, refer to the corresponding section of the UBC.

For the seismic exam, the most important parts of the UBC are

Volume 2—Structural Engineering Design Provisions

Chapter 16   Structural Design Requirements
                     Division I—General Design Requirements
                     Division II—Snow Loads
                     Division III—Wind Design
                     Division IV—Earthquake Design
                     Division V—Soil Profile Types

Chapter 18   Foundations and Retaining Walls

Chapter 19   Concrete

Chapter 20   Lightweight Metals

Chapter 21   Masonry

Chapter 22   Steel

Chapter 23   Wood

The cross-reference tables shown on the following pages are provided to assist you in locating certain key chapters, tables, figures, and formulas of the 1994 edition in the 1997 edition of the UBC. An empty block under the 1994 Reference No. column exhibits a new formula or table introduction in the 1997 edition. Another empty block under the 1997 Reference No. column implies a formula or table omitted from the 1994 edition for the 1997 edition. Asterisks next to the 1997 reference numbers indicate a change from the 1994 edition. For changes in the contents of the listed chapters, you need to refer to the 1997 UBC.

In each chapter of the UBC, you should concentrate your study on the sections that relate to earthquake regulations.

You will approach the actual exam with a high probability of success if you have taken all the recommendations of this Introduction into consideration.

Good luck on the exam!

# UBC FIGURES

| 1994 Reference No. | Description | 1997 Reference No. |
|---|---|---|
| 16-2 | Seismic Zone Map of the United States | 16-2* |
| 16-3 | Design Response Spectra | 16-3* |

# UBC FORMULAS

| 1994 Reference No. | Description | 1997 Reference No. |
|---|---|---|
| 28-1 | $V = \dfrac{2.75 Z I W}{R_w}$ | |
| 28-2 | $C = \dfrac{1.25 S}{T^{2/3}}$ | |
| | $V = \left(\dfrac{C_v I}{RT}\right) W$ | 30-4 |
| | $V = \left(\dfrac{2.5 C_a I}{R}\right) W$ | 30-5 |
| | $V = 0.11 C_a I W$ | 30-6 |
| | $V = \left(\dfrac{0.8 Z N_v I}{R}\right) W$ | 30-7 |
| 28-3 | $T = C_t (h_n)^{3/4}$ | 30-8 |
| 28-4 | $A_c = \sum A_e \left[ 0.2 + \left(\dfrac{D_e}{h_n}\right)^2 \right]$ | 30-9 |
| 28-5 | $T = 2\pi \sqrt{\dfrac{\sum\limits_{i=1}^{n} w_i \delta_i^2}{g \sum\limits_{i=1}^{n} f_i \delta_i}}$ | 30-10 |
| | $V = \left(\dfrac{3.0 C_a}{R}\right) W$ | 30-11 |
| | $F_x = \left(\dfrac{3.0 C_a}{R}\right) W_i$ | 30-12 |

# UBC FORMULAS (continued)

| 1994 Reference No. | Description | 1997 Reference No. |
|---|---|---|
| 28-6 | $V = F_t + \sum_{i=1}^{n} F_i$ | 30-13 |
| 28-7 | $F_t = 0.07TV$ | 30-14 |
| 28-8 | $F_x = \dfrac{(V - F_t)w_x h_x}{\sum_{i=1}^{n} w_i h_i}$ | 30-15 |
| | $\Delta_M = 0.7R\Delta_s$ | 30-17 |
| 30-1 | $F_p = ZI_p C_p W_p$ | |
| | $F_p = 4.0 C_a I_p W_p$ | 32-1 |
| | $F_p = \left(\dfrac{a_p C_a I_p}{R_p}\right)\left(1 + (3)\left(\dfrac{h_x}{h_r}\right)\right) W_p$ | 32-2 |
| 31-1 | $F_{px} = \left(\dfrac{F_t + \sum_{i=x}^{n} F_i}{\sum_{i=x}^{n} w_i}\right) w_{px}$ | 33-1 |
| | $\Delta_{MT} = \sqrt{(\Delta_{M1})^2 + (\Delta_{M2})^2}$ | 33-2 |
| 32-1 | $V = 0.5 ZIW$ | |
| | $V = 0.7 C_a IW$ | 34-1 |
| | $V = 0.56 C_a IW$ | 34-2 |
| | $V = \left(\dfrac{1.6 Z N_v I}{R}\right) W$ | 34-3 |
| | $\overline{\nu}_s = \dfrac{\sum_{i=1}^{n} d_i}{\sum_{i=1}^{n} \dfrac{d_i}{\nu_{si}}}$ | 36-1 |

# UBC CHAPTERS

| 1994 Reference No. | Description | 1997 Reference No. |
|---|---|---|
| 16 | Structural Forces | 16 |
| 21 | Masonry | 21 |
| 23 | Wood | 23 |
| 19 | Concrete | 19 |
| 22 | Steel | 22 |
| 18 | Foundations and Retaining Walls | 18 |

# UBC TABLES

| 1994 Reference No. | Description | 1997 Reference No. |
|---|---|---|
| 16-I | Seismic Zone Factor, $Z$ | 16-I |
| 16-J | Soil Profile Types | 16-J* |
| 16-K | Occupancy Category | 16-K |
| 16-L | Vertical Structural Irregularities | 16-L* |
| 16-M | Plan Structural Irregularities | 16-M* |
| 16-N | Structural Systems | 16-N* |
| 16-O | Horizontal Force Factors, $a_p$ and $R_p$ | 16-O* |
| 16-P | $R$ and $\Omega_o$ Factors for Nonbuilding Structures | 16-P* |
| | Seismic Coefficient, $C_a$ | 16-Q |
| | Seismic Coefficient, $C_v$ | 16-R |
| | Near-Source Factor, $N_a$ | 16-S |
| | Near-Source Factor, $N_v$ | 16-T |
| | Seismic Source Type | 16-U |
| 21-N | Allowable Shear on Bolts for Empirically Designed Masonry Except Unburned Clay Units | 21-N |
| 23-I-F | Bolt Design Values (Z) for Single Shear (Two Member) Connections | 23-III-B-1* |
| | Bolt Design Values (Z) for Double Shear (Three Member) Connections | 23-III-B-2* |
| 23-I-G | Box Nail Design Values (Z) for Single Shear (Two Member) Connections | 23-III-C-1* |
| | Common Wire Nail Design Values (Z) for Single Shear (Two Member) Connections | 23-III-C-2* |

# UBC TABLES (continued)

| 1994 Reference No. | Description | 1997 Reference No. |
|---|---|---|
| 23-I-I | Maximum Diaphragm Dimension Ratios | 23-II-G* |
| 23-I-J-1 | Allowable Shear in Pounds per Foot for Horizontal Wood Structural Panel Diaphragms with Framing of Douglas Fir-Larch or Southern Pine | 23-II-H |
| 23-I-K-1 | Allowable Shear for Wind or Seismic Forces in Pounds per Foot for Wood Structural Panel Shearwalls with Framing of Douglas Fir-Larch or Southern Pine | 23-II-I-1 |

# NOMENCLATURE

The following list presents symbols and their corresponding units.

| | | | | |
|---|---|---|---|---|
| $a$ | acceleration (ft/sec$^2$, m/s$^2$) | | $E_v$ | earthquake load |
| $a_p$ | in-structure component amplification factor | | $f$ | frequency (Hz) |
| $A$ | area (ft$^2$, m$^2$) | | $f_y$ | allowable tensile steel (lbf/in$^2$, kPa) |
| $b$ | width of a horizontal diaphragm (ft, m) | | $F$ | force or load (lbf, N) |
| $B$ | damping coefficient (lbf-sec/ft, N·s/m) | | $F_s$ | stress in reinforcing steel (lbf/in$^2$, kPa) |
| $C$ | chord force (lbf, N) | | $g$ | acceleration of gravity (ft/sec$^2$, m/s$^2$) |
| $C_a$ | seismic response coefficient | | $G$ | shear modulus (lbf/ft$^2$, kPa) |
| $C_M$ | center of mass (ft, m) | | $h$ | height (ft, m) |
| $C_p$ | horizontal force factor | | $h_n$ | height above the base to the $n^{\text{th}}$ level (ft, m) |
| $C_R$ | center of rigidity (ft, m) | | $h_r$ | structure roof elevation with respect to grade |
| $C_t$ | numerical coefficient | | | |
| $C_v$ | seismic response coefficient | | $h_x$ | element or component attachment elevation with respect to grade |
| $D$ | dead load (lbf, N) | | | |
| $D$ | drag strut force (lbf, N) | | $I$ | seismic importance factor |
| $e$ | eccentricity (ft, m) | | $I$ | moment of inertia (ft$^4$, m$^4$) |
| $E$ | earthquake load | | | |
| $E$ | modulus of elasticity (lbf/in$^2$, kPa) | | $I_p$ | seismic importance factor |
| $E_h$ | earthquake load | | $k$ | stiffness (lbf/ft, N/m) |
| $E_m$ | earthquake load | | $L$ | length (ft, m) |

$L$     live load (lbf, N)

$m$     mass (lbm, kg)

$M$     magnitude

$M$     moment (ft-lbf, N·m)

$n$     exponent

$n$     modular ratio

$N_a$     near source factor

$N_v$     near source factor

$\rho$     reliability/redundancy factor

$PI$     plasticity index of soil

$R$     reaction force (lbf, N)

$R$     relative rigidity

$R$     response modification factor

$R_p$     component response modification factor

$s$     distance (ft, m)

$S$     seismic site/soil coefficient

$S$     story strength

$S_a$     spectral acceleration (ft/sec$^2$, m/s$^2$)

$S_A$     soil profile type

$S_B$     soil profile type

$S_C$     soil profile type

$S_d$     spectral displacement (ft, m)

$S_D$     soil profile type

$S_E$     soil profile type

$S_F$     soil profile type

$\overline{S}_u$     undrained shear strength (lbf/ft$^2$, kPa)

$S_v$     spectral velocity (ft/sec, m/s)

$t$     thickness (ft, m)

$T$     elastic fundamental period of vibration (sec, s)

$T$     torsional moment (ft-k, kN·m)

$v$     shear stress (lbf/ft$^2$, kPa)

$V$     base shear (lbf, N)

$V$     shear in horizontal diaphragm or shear wall (lbf, N)

$W$     lateral distributed force on horizontal diaphragm (lbf, N)

$W$     weight (lbf, N)

$W_i$     dead load tributary to the $i_{\mathrm{th}}$ floor (lbf, N)

$W_r$     dead load tributary to the roof (lbf, N)

$W_{px}$     weight of diaphragm and elements tributary to level $x$ (lbf, N)

$W_x$     dead load tributary to the roof (lbf, N)

$x$     drift (ft, m)

$\overline{x}$     center of mass in $x$-direction (ft, m)

$\overline{y}$     center of mass in $y$-direction (ft, m)

$Z$     seismic zone factor

$\Delta$     deflection (in, cm)

$\Delta$     drift (in, cm)

$\Delta_M$     maximum inelastic response displacement

$\Delta_S$     design level response displacement

$\epsilon$     strain

$\xi$     damping ratio

$\sigma$     normal stress (lbf/in$^2$, kPa)

$\vartheta$     unit shear in horizontal diaphragm or shear wall (lbf/ft, N/m)

$\omega$     angular natural frequency (rad/sec, rad/s)

$w$     uniform load per unit length (lbf/ft, N/m)

$\Omega_o$     seismic force amplification factor

$\rho$     reliability/redundancy factor

# SUBSCRIPTS

$a$     accidental

$c$     critical

$i$     level number

$n$     uppermost level in the main portion of the structure

$p$     part or portion

$r$     roof

$s$     steel

$t$     top

$x$     level number

$x$     with respect to $x$-axis

$y$     with respect to $y$-axis

# CHAPTER 1

## SEISMOLOGY PRINCIPLES, EARTHQUAKE CHARACTERISTICS, AND BASIC STRUCTURAL DYNAMICS

### TOPICS

Acceleration

Amplification of Motion

Attenuation

Base Shear

Capable Earthquake

Compression Wave

Damping

Damping Period

Earthquake Fault

Epicenter

Flexibility

Ground Motion

Intensity of Earthquake

Location of Earthquake

Lumped Mass

Magnitude of Earthquake

Maximum Peak Ground Acceleration

Mode

Modified Mercalli Intensity Scale

Natural Frequency

Natural Period

Oceanic Plate

Probability of Occurrence

Probable Earthquake

P-Wave

Resonance

Richter Magnitude Scale

Rigidity

Seismic Sea Wave

Seismometer

Shear Wave

Single-Degree-of-Freedom System

Soil Liquefaction

Spectral Velocity

Stiffness

Subduction

S-Wave

Tsunami

1. Which of the following terms describe the location of an earthquake?

      I. epicenter
      II. focal depth
      III. dip angle

  A. I
  B. II
  C. I and II
  D. I and III

2. There is a footwall facing a hanging wall prior to an earthquake. The hanging wall moves up during the earthquake. Which type of fault occurred?

  A. strike-slip fault
  B. reverse fault
  C. normal fault
  D. oblique fault

3. The San Andreas fault of California is which type of fault?

  A. right-lateral fault
  B. left-lateral fault
  C. normal fault
  D. reverse fault

4. The process where an oceanic plate slides beneath a continental plate is known as

  A. lithosphere.
  B. mantle.
  C. subduction.
  D. strike-slip.

5. Which of the following statements is incorrect?

  A. Compression, shear, and surface waves are seismic waves.
  B. S-waves are at right angles to the compression waves.
  C. S-waves can be expressed as horizontal and vertical components.
  D. None of the above are incorrect.

6. Which of the following statements is incorrect?

  A. S-waves travel more slowly than P-waves.
  B. S-waves transmit more energy than P-waves.
  C. S-waves cause less damage to structures than P-waves.
  D. All of the above are incorrect.

7. Which of the following activities occurs as seismic sea waves approach land?

  A. Wave velocity increases.
  B. Wave height decreases.
  C. Both wave velocity and height increase.
  D. Wave velocity decreases and wave height increases.

8. How are seismic waves generated?

      I. by volcanic eruptions
      II. by deep artificially induced explosions
      III. by sudden dislocations within the earth's crust

  A. I and II
  B. II and III
  C. I and III
  D. I, II, and III

9. What is a tsunami?

  A. a seismic sea wave
  B. a tidal wave
  C. a surface-water wave
  D. all of the above

10. What does a seismometer measure?

  A. components of ground motion
  B. reference points
  C. attenuation
  D. actual displacement

11. Which of the following statements are correct?

    I. The Richter scale measures earthquake strength.

    II. The magnitude of an earthquake can be determined from the logarithm of the recorded amplitude.

    III. The magnitude of an earthquake does not depend on the length of the fault slip.

    A. I and II
    B. I and III
    C. II and III
    D. I, II, and III

12. What does the Modified Mercalli Intensity scale measure?

    I. magnitude
    II. apparent severity at particular location
    III. intensity

    A. I and II
    B. II and III
    C. I and III
    D. I, II, and III

13. How many levels of intensity are associated with the Modified Mercalli Intensity scale?

    A. 8
    B. 10
    C. 12
    D. 14

14. What is the definition of the term *attenuation*?

    A. ground motion
    B. a decrease in seismic energy
    C. a geologic formation
    D. the orientation of a fault

15. Attenuation is not influenced by

    A. path line and length.
    B. geologic formation.
    C. focal depth.
    D. earthquake magnitude.

16. Which of the following periods share the same definition?

    I. the period of an earthquake
    II. the period of a site
    III. the period of a building

    A. I and II
    B. I and III
    C. II and III
    D. none of the above

17. Which of the following characteristics influence the amount of structural damage caused by an earthquake?

    I. peak ground acceleration and duration of motion

    II. soil condition at the site and period of the site

    III. distance between the epicenter and the structure

    IV. natural period and damping of the structure

    A. I and II
    B. III and IV
    C. I, II, and III
    D. all of the above

18. When does resonance occur?

    I. when the building period coincides with the earthquake period

    II. when the earthquake period coincides with the site period

    III. when the site period coincides with the building period

    A. I and II
    B. I and III
    C. II and III
    D. I, II, and III

19. Permanent, heavy air-conditioning equipment is installed on top of a building. How will this affect the fundamental period of the building?

    A. It will increase.
    B. It will decrease.
    C. It will remain the same.
    D. It will increase the acceleration.

20. With which of the following is an increase in earthquake magnitude correlated?

    A. an increase in acceleration
    B. an increase in duration
    C. a decrease in acceleration and duration
    D. magnitude, acceleration, and duration are not necessarily correlated

21. Which of the following characteristics influence the effect of an earthquake?

    A. frequency
    B. duration
    C. ground acceleration
    D. all of the above

22. Liquefaction is best described as

    A. a sudden drop in the shear strength of a soil.
    B. a decrease in pore water pressure of a soil.
    C. an increase in the bearing capacity of a soil.
    D. an increase in the effective stress of a soil.

23. When liquefaction occurs, which of the following soil conditions are likely to exist?

    I. The soils have zero shear strength.
    II. The soils are made of saturated, cohesionless particles.
    III. The soils are made of cohesion particles.

    A. II only
    B. I and III
    C. III only
    D. I and II

24. Which of the following sites is most likely to amplify ground motion?

    A. a dense soil site
    B. a rock site
    C. a stiff clay site
    D. a soft clay site

25. In the vicinity of the epicenters of major California earthquakes, what is the approximate peak ground acceleration?

    A. 0.10 g
    B. 0.25 g
    C. 0.50 g
    D. 1.25 g

26. Spectral velocity is best described as

    A. the velocity of seismic energy.
    B. the velocity of a secondary wave.
    C. the velocity of a structure.
    D. the velocity of a structure relative to the ground.

27. For a single-degree-of-freedom system, the natural period, $T$, is

    I. the inverse of the natural frequency.
    II. equivalent to the linear natural frequency.
    III. expressed in Hz.

    A. I
    B. II
    C. III
    D. I and III

28. What is the spectral acceleration of a single-degree-of-freedom system?

    A. It is the minimum acceleration experienced by the system in response to a perturbation.
    B. It is the average acceleration experienced by the system in response to a perturbation.
    C. It is the maximum acceleration experienced by the system in response to a perturbation.
    D. It is the absolute acceleration experienced by the system in response to a perturbation.

29. For the following harmonic oscillator, how is the stiffness best described?

    A. as the force acting on the ideal linear spring
    B. as the force deflecting the spring a distance of 1 unit
    C. as the magnitude of the spring deflection
    D. as the reciprocal of deflection

30. When does the natural building period coincide with the earthquake period?

    A. when the natural frequency is at its maximum
    B. when the acceleration is at its maximum
    C. when the displacement is at its maximum
    D. all of the above

31. The period of a structure is correlated with the

    I. mass of the structure.
    II. stiffness of the structure.

    A. I only
    B. II only
    C. I and II
    D. none of the above

32. What is the maximum considered (capable) earthquake motion at a site?

    A. It is the motion intensity with a 10% probability of being exceeded in a 100-year time period.
    B. It is based on presently available data and knowledge of the site.
    C. It is the maximum level of earthquake ground motion expected at the site.
    D. All of the above are true.

33. Consider the natural period, $T$, and acceleration, $a$, of a single-degree-of-freedom system. When the mass, $m$, of the system increases, how are $T$ and $a$ affected?

    A. $T$ increases and $a$ increases.
    B. $T$ increases and $a$ decreases.
    C. $T$ decreases and $a$ increases.
    D. $T$ decreases and $a$ decreases.

34. Where $K$ is the stiffness and $\Delta$ is the deflection, which of the following relationships best describes the rigidity, $R$?

    A. $\dfrac{1}{K}$
    B. $K\Delta$
    C. $\Delta$
    D. $\dfrac{1}{\Delta}$

35. A mass hangs on two ideal springs as shown. What is the total composite spring constant for the system?

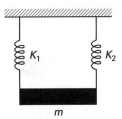

    A. $K_1 + K_2$
    B. $K_1 K_2$
    C. $\dfrac{1}{K_1} + \dfrac{1}{K_2}$
    D. $\dfrac{K_1 K_2}{K_1 + K_2}$

36. Each of the following pendulums (equal in mass) are hanging on an ideal spring. The periods of vibration for the pendulums are 1.73 sec (1.73 s) and 3.0 sec (3.0 s), respectively. Assume the pendulums are on frictionless pivots. What is the stiffness of the second pendulum's spring?

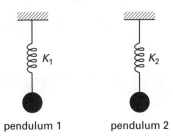

    pendulum 1    pendulum 2

    A. $\dfrac{1}{3}K_1$
    B. $\dfrac{1}{2}K_1$
    C. $K_1$
    D. 1.33 $K_1$

37. Each of the following columns supports a block of identical mass, $m$. The columns are fixed at the bottom and free at the top. The height of the second column is twice the height of the first column. The modulus of elasticity, $E$, and moment of inertia, $I$, for both

columns are the same. The systems have natural periods of vibration of $T_1$ and $T_2$, respectively. Neglecting the weight of the columns, what is the natural period of vibration for the second system?

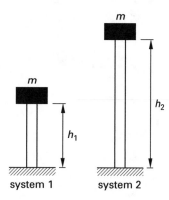

A. $0.5T_1$

B. $2T_1$

C. $3T_1$

D. $4T_1$

38. A force is acting at the top of a building frame as shown. The supporting columns are of equal height and are fixed at the base. The modulus of elasticity, $E$, is the same for each column. Assuming the top plate is rigid and $I_1 = \frac{1}{3}I_2 = \frac{1}{6}I_3$, what is the shear distributed to the first column?

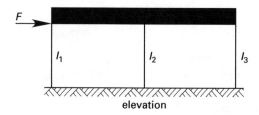

elevation

A. $\dfrac{1}{10}F$

B. $\dfrac{1}{5}F$

C. $\dfrac{2}{3}F$

D. $\dfrac{1}{3}F$

39. How is the term *damping* described?

    I. the ratio of one cycle's amplitude to the subsequent cycle

    II. the dynamic magnification factor

    III. the dissipation of energy from an oscillating system

A. I only

B. III only

C. I and III

D. II and III

40. How is the term *flexibility* defined?

A. stiffness

B. the reciprocal of stiffness

C. rigidity

D. static deflection

41. From the following illustrated spectra, which curve represents a moderate damping?

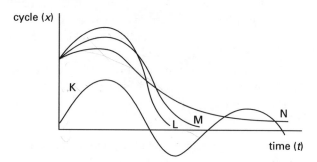

A. K

B. L

C. M

D. N

42. For a system with several masses and many modes, it is true that

    I. the characteristic shape of each mode is unique.

    II. all modes have the same natural frequency.

    III. the fundamental mode has the longest period.

A. I only

B. II only

C. I and III

D. II and III

43. For a system with lumped masses and many modes, which of the following statements is correct?

    A.  Higher modes have lower frequencies.
    B.  Higher modes have higher frequencies.
    C.  Higher modes have longer periods.
    D.  Lower modes have shorter periods.

44. Compared with buildings with few stories, high-rise buildings have which of the following?

    A.  higher frequencies
    B.  longer periods
    C.  higher acceleration
    D.  higher stiffness

45. Consider the ratio of building acceleration to ground acceleration. Which of the following statements is incorrect?

    A.  The ratio depends on the building period.
    B.  The ratio is equal to 1.0 for an infinitely stiff building.
    C.  The ratio is equal to 1.0 for buildings with zero natural period.
    D.  Building acceleration is typically lower than ground acceleration.

46. In the response spectra shown, what is the base shear? Assume $W = 160$ k (712 kN) and $T = 0.3$ sec.

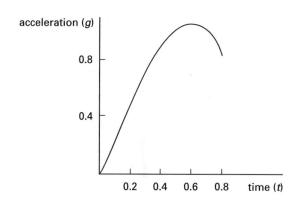

    A.  40 k   (180 kN)
    B.  70 k   (320 kN)
    C.  110 k   (500 kN)
    D.  140 k   (640 kN)

**1. Answer C**

The point on the earth's surface directly above the focus is the *epicenter*, and the depth from the earth's surface to the focus is the *focal depth*. Therefore, the epicenter and the focal depth describe the location of an earthquake.

**2. Answer B**

Since the hanging wall is thrust upward and over to the footwall, it is a reverse (thrust) fault.

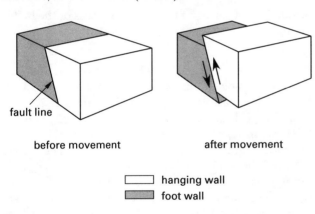

fault line

before movement          after movement

☐ hanging wall
▨ foot wall

**3. Answer A**

The sudden displacement, or *slip*, of the crust or rock along the San Andreas fault of California is right-lateral movement. When a person stands on either side of a right-lateral fault and looks across the fault, the fault movement will be to that person's right.

**4. Answer C**

The term *subduction* refers to the process where an oceanic plate drops and slides beneath a continental plate.

**5. Answer D**

The seismic waves are compression, shear, and surface waves. From the epicenter, the compression waves (primary waves or P-waves) travel through the earth's interior to reach the surface first. Compression waves displace materials directly behind or ahead of their path of travel, while shear waves (secondary waves or S-waves) displace material at right angles to their line of travel and reach the surface later. The S-waves have horizontal and vertical components since their propagation path may be in any direction from the source.

**6. Answer C**

S-waves travel more slowly and reach the surface after the P-waves. They transmit more energy, thus causing the bulk of damage to structures.

**7. Answer D**

During an earthquake, sudden changes in sea floor depth set a massive wave in motion. As the wave approaches land from a deep sea floor to a shallow sea floor, the velocity decreases and the wave height increases due to increased friction with the shallow sea floor.

**8. Answer D**

Sudden dislocations and changes within the earth's crust create seismic waves. Since volcanic eruptions and deep artificially induced explosions result in displacement, they also generate seismic waves.

**9. Answer D**

In Japanese, a seismic sea wave is called a *tsunami*. It is also known as a *tidal wave* or *surface-water wave*.

**10. Answer D**

A seismometer is a seismic instrument that measures the actual displacement of the ground with respect to a fixed reference point. The magnitude, $M$, of the earthquake can be calculated from the logarithm of the amplitude of the displacement.

**11. Answer A**

The Richter scale measures earthquake strength. From the logarithm of the recorded amplitude, the magnitude, $M$, can be determined. The magnitude is generally related to the length of the fault slip. As the length of the vertical or horizontal fault slip increases, the earthquake magnitude tends to increase.

**12. Answer B**

The Modified Mercalli Intensity scale measures the intensity of the observed effects and the damage caused by an earthquake.

**13. Answer C**

The Modified Mercalli Intensity scale consists of 12 levels of intensity. The levels are expressed as Roman numerals. These levels range from unapparent shaking to total destruction and devastation.

**14. Answer B**

The amount of energy in a seismic wave decreases when it propagates through rock or other geologic material. This decrease of seismic energy with distance from the source is called *attenuation*.

**15. Answer D**

Attenuation can be described as a decrease in seismic energy. The factors influencing attenuation are path line, path length, geologic formations, focal depth, location of the epicenter, and the properties of the rock. The magnitude of an earthquake does not decrease the amount of its energy.

**16. Answer D**

The period of the earthquake refers to the predominant period of the wave and can be determined by Fourier analysis. The site (soil) and the building both have their own fundamental or natural periods. The site (soil) period is determined from geotechnical data, while the building period is determined by an analysis of the structure.

**17. Answer D**

The degree of the structural damage due to an earthquake depends on the ground acceleration, duration of motion, frequency content, local soil conditions, period of the site, distance between the epicenter and the structure, geologic formations between the epicenter and the structure, natural period and damping of the structure, type of materials used for the structure, and seismic provisions considered at the time of building design and construction.

**18. Answer D**

Resonance results in amplification of response and occurs where the building, earthquake, and/or site (soil) periods coincide with each other. An example where resonance between all these components occurred was the 1985 Mexico City earthquake. By the time the earthquake reached the city from its epicenter 365 km away, the acceleration amplitude was small but the predominant period matched the lake bed soils beneath the city. This resulted in resonance and magnification of the ground motions. In addition, some buildings with the same natural period as the ground motion experienced a magnified response.

**19. Answer A**

$$T = 2\pi \sqrt{\frac{m}{K}} \qquad \text{[SI]}$$

$$T = 2\pi \sqrt{\frac{m}{g_c K}} \qquad \text{[U.S.]}$$

The stiffness of the building will not be affected by the installation of the equipment. As the mass of the building increases, the fundamental period also increases.

**20. Answer D**

The magnitude of an earthquake has no correlation with the acceleration or duration. For example, the 1989 Loma Prieta earthquake with a magnitude of 7.1 registered a peak ground acceleration of 0.65 g while the 1994 Northridge earthquake with a magnitude of 6.7 recorded a ground acceleration of 1.8 g. The strong phase of the 1971 San Fernando earthquake (6.6 magnitude) lasted only 7 sec, while for the 1940 El Centro earthquake (6.4 magnitude), the duration of strong acceleration lasted 16 sec.

**21. Answer D**

The frequency, duration, and ground acceleration affect the amount of seismic and structural damage.

**22. Answer A**

Liquefaction occurs in soils. A soil behaves like a liquid when its shear strength drops suddenly. Liquefaction implies an increase in pore water pressure, a complete loss of bearing capacity, and a decrease in the effective stress of a soil.

**23. Answer D**

Liquefaction is a phenomenon in soils in which the shear strength is drastically reduced due to the sudden application of an earthquake force. This occurs most often in soils that are saturated in a loose condition and made of cohesionless particles, such as sand.

**24. Answer D**
Earthquakes may trigger rapid consolidation of soft clay, whereby the loss of grain-to-grain contact and presence of excess pore water leads to complete loss of strength. Soft clay sites tend to increase the amplitude of earthquake motions more than other local soil conditions.

**25. Answer C**
Based on the ground accelerations recorded during earthquakes in California, generally speaking, for earthquakes with a magnitude between 8.0 and 8.5, the ground acceleration is 0.5 g.

**26. Answer D**
The maximum vibration of a single-degree-of-freedom system is measured in terms of acceleration, velocity, or displacement. The maximum velocity of a structure relative to the ground is known as the *spectral velocity.*

**27. Answer A**
For a single-degree-of-freedom system, the natural period, $T$, is the time in which the system completes one cycle of oscillation. The natural period is the inverse of the natural frequency, which can be expressed as $T = 1/f$.

**28. Answer C**
The maximum vibration of a single-degree-of-freedom system is measured in terms of acceleration, velocity, or displacement. The maximum acceleration experienced by the system in response to a disturbing force is known as the *spectral acceleration.*

**29. Answer B**
Hooke's law relates the amount of deflection of an ideal linear spring to the force applied to it. It is expressed as $F = Kx$. The stiffness, $K$, is the force that deflects the spring a distance of 1 unit.

**30. Answer B**
When a building period coincides with an earthquake period, resonance occurs and the acceleration response is at its maximum.

**31. Answer C**

$$T = 2\pi\sqrt{\frac{m}{K}} \quad \text{[SI]}$$

$$T = 2\pi\sqrt{\frac{m}{g_cK}} \quad \text{[U.S.]}$$

As the weight of a structure increases, the period of vibration becomes longer. For example, steel or concrete structures tend to have high periods compared with wood structures. The stiffness of a structure also affects the period of vibration: the stiffer the structure, the shorter the period of vibration.

**32. Answer D**
The maximum considered (capable) earthquake is defined in the UBC [Sec. 1655, Appendix, DIV IV] as the maximum level of earthquake ground shaking that may be expected at the building site within the known geological framework. In seismic zones 3 and 4, this intensity may be taken as the level of earthquake ground motion that has a 10% probability of being exceeded in 100 years.

**33. Answer B**
The mass and period of a system are related by the following equation.

$$T = 2\pi\sqrt{\frac{m}{K}} \quad \text{[SI]}$$

$$T = 2\pi\sqrt{\frac{m}{g_cK}} \quad \text{[U.S.]}$$

Increasing the mass of a single-degree-of-freedom system will result in a longer period of vibration. From Newton's second law, $F = ma$, it is seen that the response of the larger mass to the same force will result in a lower acceleration.

**34. Answer D**
Rigidity is the reciprocal of deflection.

**35. Answer A**
For parallel springs, the composite stiffness is the sum of the individual spring stiffnesses.

**36. Answer A**

*SI solution*

The natural period of vibration for a single-degree-of-freedom system is

$$T_1 = 2\pi\sqrt{\frac{m_1}{K_1}}$$

$$T_1^2 = 4\pi^2\left(\frac{m_1}{K_1}\right)$$

$$T_1^2 K_1 = 4\pi^2 m_1$$

$$T_2 = 2\pi\sqrt{\frac{m_2}{K_2}}$$

$$T_2^2 = 4\pi^2\left(\frac{m_2}{K_2}\right)$$

$$T_2^2 K_2 = 4\pi^2 m_2$$

Since $m_1 = m_2$,

$$T_1^2 K_1 = T_2^2 K_2$$
$$(1.73)^2 K_1 = (3.0)^2 K_2$$
$$K_2 = \tfrac{1}{3} K_1$$

*Customary U.S. solution*

The natural period of vibration for a single-degree-of-freedom system is

$$T_1 = 2\pi\sqrt{\frac{m_1}{g_c K_1}}$$

$$T_1^2 = 4\pi^2\left(\frac{m_1}{g_c K_1}\right)$$

$$T_1^2 g_c K_1 = 4\pi^2 m_1$$

$$T_2 = 2\pi\sqrt{\frac{m_2}{g_c K_2}}$$

$$T_2^2 = 4\pi^2\left(\frac{m_2}{g_c K_2}\right)$$

$$T_2^2 g_c K_2 = 4\pi^2 m_2$$

Since $m_1 = m_2$,

$$T_1^2 g_c K_1 = T_2^2 g_c K_2$$
$$T_1^2 K_1 = T_2^2 K_2$$
$$(1.73)^2 K_1 = (3.0)^2 K_2$$
$$K_2 = \tfrac{1}{3} K_1$$

**37. Answer C**

*SI solution*

Both systems can be modeled as cantilever beams because they are fixed at the bottom and free at the top. The stiffness of a cantilever beam is described by $K = 3EI/h^3$.

$$K_1 = \frac{3E_1 I_1}{h_1^3}$$

$$3E_1 I_1 = K_1 h_1^3$$

$$K_2 = \frac{3E_2 I_2}{h_2^3}$$

$$3E_2 I_2 = K_2 h_2^3$$

It is given that $h_2 = 2h_1, E_1 = E_2$, and $I_1 = I_2$. Therefore,

$$K_1 h_1^3 = K_2 h_2^3$$
$$K_1 h_1^3 = K_2 (2h_1)^3$$
$$K_1 = 8K_2 \qquad \text{[Eq. 1]}$$

$$T_1 = 2\pi\sqrt{\frac{m_1}{K_1}}$$

$$T_1^2 = 4\pi^2\left(\frac{m_1}{K_1}\right)$$

$$T_1^2 K_1 = 4\pi^2 m_1$$

$$T_2 = 2\pi\sqrt{\frac{m_2}{K_2}}$$

$$T_2^2 = 4\pi^2\left(\frac{m_2}{K_2}\right)$$

$$T_2^2 K_2 = 4\pi^2 m_2$$

Since $m_1 = m_2$,

$$T_1^2 K_1 = T_2^2 K_2 \qquad \text{[Eq. 2]}$$

Substitute Eq. 1 into Eq. 2 and solve.

$$T_1^2(8K_2) = T_2^2(K_2)$$
$$T_2^2 = 8T_1^2$$
$$T_2 = 2.83 T_1$$
$$\approx 3.0 T_1$$

*Customary U.S. solution*

Both systems can be modeled as cantilever beams because they are fixed at the bottom and free at the top. The stiffness of a cantilever beam is described by $K = 3EI/h^3$.

$$K_1 = \frac{3E_1 I_1}{h_1^3}$$

$$3E_1 I_1 = K_1 h_1^3$$

$$K_2 = \frac{3E_2 I_2}{h_1^3}$$

$$3E_2 I_2 = K_2 h_2^3$$

It is given that $h_2 = 2h_1$, $E_1 = E_2$, and $I_1 = I_2$. Therefore,

$$K_1 h_1^3 = K_2 h_2^3$$
$$K_1 h_1^3 = K_2 (2h_1)^3$$
$$K_1 = 8K_2 \qquad \text{[Eq. 1]}$$

$$T_1 = 2\pi \sqrt{\frac{m_1}{g_c K_1}}$$

$$T_1^2 = 4\pi^2 \left(\frac{m_1}{g_c K_1}\right)$$

$$T_1^2 g_c K_1 = 4\pi^2 m_1$$

$$T_2 = 2\pi \sqrt{\frac{m_2}{g_c K_2}}$$

$$T_2^2 = 4\pi^2 \left(\frac{m_2}{g_c K_2}\right)$$

$$T_2^2 g_c K_2 = 4\pi^2 m_2$$

Since $m_1 = m_2$,

$$T_1^2 K_1 = T_2^2 K_2 \qquad \text{[Eq. 2]}$$

Substitute Eq. 1 into Eq. 2 and solve.

$$T_1^2 (8K_2) = T_2^2 (K_2)$$
$$T_2^2 = 8T_1^2$$
$$T_2 = 2.83 T_1$$
$$\approx 3.0 T_1$$

38. Answer A

Since the columns are rigid, the force is distributed to the columns in proportion to their rigidities (or stiffnesses). The composite stiffness is the sum of the individual column stiffnesses.

The distributed shear stress in the first column is

$$V_1 = F \left(\frac{K_1}{K_1 + K_2 + K_3}\right) \qquad \text{[Eq. 1]}$$

Substitute for $K_1$, $K_2$, and $K_3$; then simplify.

$$K_1 = \frac{12E_1 I_1}{h_1^3}$$

$$K_2 = \frac{12E_2 I_2}{h_2^3}$$

$$K_3 = \frac{12E_3 I_3}{h_3^3}$$

It is given that $E_1 = E_2 = E_3$ and $h_1 = h_2 = h_3$. Therefore,

$$K_T = K_1 + K_2 + K_3$$
$$= \left(\frac{12E_1}{h_1^3}\right)(I_1 + I_2 + I_3) \qquad \text{[Eq. 2]}$$

Substitute Eq. 2 into Eq. 1 and solve.

$$V_1 = F \left(\frac{I_1}{I_1 + I_2 + I_3}\right) = F \left(\frac{I_1}{I_\text{total}}\right)$$

For $I_1 = \frac{1}{3} I_2$ and $I_2 = \frac{1}{6} I_3$

$$I_\text{total} = I_1 + 3I_1 + 6I_1 = 10I_1$$

Therefore,

$$V_1 = F \left(\frac{I_1}{10I_1}\right) = \frac{1}{10} F$$

39. Answer B

When a system is set into oscillatory motion, it will continuously move until the dissipation of energy (primarily through friction) eventually causes the system to reach a motionless equilibrium position. This dissipation of energy is called *damping*.

40. Answer B

*Flexibility* is the total deflection of a structural system when a unit lateral force is applied. Flexibility can also be defined as the inverse of stiffness.

41. Answer A

Curve K represents an oscillating system that takes many cycles (and a long time) before reaching the motionless equilibrium position. Systems with small or moderate amounts of damping are called *underdamped*; curves L, M, and N are *critically damped* or *overdamped*.

### 42. Answer C

For a multiple-degree-of-freedom system, the oscillation of the system is a combination of the oscillations of the several lumped masses. Each mode of oscillation (as many modes as there are lumped masses) has its own shape and natural frequency. The first, or fundamental, mode has the longest period (i.e., the smallest frequency).

### 43. Answer B

For a multiple-degree-of-freedom system, at higher modes the periods are smaller (i.e., frequencies are longest). The lower modes have the longest period (i.e., the lowest frequency).

### 44. Answer B

High-rise buildings require greater flexibility to resist large earthquakes. In general, the height of these buildings influences the period of vibration—the taller the high-rise, the longer the period of vibration.

### 45. Answer D

The ground acceleration is an acceleration to which a building responds, while the building acceleration is a function of its dynamic characteristics, mass, $m$, and stiffness, $K$. The response of the building to the ground acceleration depends on the characteristics of both the earthquake and the structure, including the building period. For infinitely stiff buildings and buildings with zero natural periods, the ground acceleration and the building acceleration are identical. Therefore, this ratio becomes 1. Generally, the building acceleration is higher than the ground acceleration.

### 46. Answer C

*SI solution*

Using the response spectra provided, the spectral acceleration for a natural period of 0.3 sec is determined to be 0.7 g.

$$F = ma$$
$$= \left(\frac{W}{g}\right) a$$
$$= \left(\frac{W}{g}\right)(0.7\ g)$$
$$= 0.7W$$
$$= (0.7)(712\ \text{kN})$$
$$= 498\ \text{kN}$$
$$\approx 500\ \text{kN}$$

*Customary U.S. solution*

Using the response spectra provided, the spectral acceleration for a natural period of 0.3 sec is determined to be 0.7 g.

$$F = \left(\frac{W g_c}{g}\right) a$$
$$= \frac{W\left(32.2\ \dfrac{\text{ft}}{\text{sec}^2\text{-g}}\right)(0.7\ g)}{32.2\ \dfrac{\text{ft-lbm}}{\text{lbf-sec}^2}}$$
$$= 0.7W\ \text{lbf}$$
$$= (0.7)(160\ \text{k})\left(\frac{1000\ \text{lbf}}{1\ \text{k}}\right)$$
$$= 112{,}000\ \text{lbf}$$
$$\approx 110\ \text{k}$$

# CHAPTER 2
## Codes and Regulatory Provisions

### TOPICS

Allowable Soil Bearing Value

Allowable Stress

Base Shear

Bearing Wall System

Braced Frame Structure

Building Frame Structure

Building Period

$C$ Coefficient

Combination Load

Component Response Modification Factor

Concentric Braced Frame

Deformation Compatibility

Dual System

Ductility

Eccentric Braced Frame

Emergency Facility

Energy Absorption Capacity

Essential Facility

Flexible Structure

Floor Partition Load

Geometric Irregularity

High-Rise Building

$h_n$ Definition

Importance Factor

Inherent Overstrength and Global Ductility Capacity

In-Structure Component Amplification Factor

Intermediate Moment-Resisting Space Frame

Interstory Displacement

Lateral Force Resisting System

Live Load

Mass Story

Maximum Inelastic Response Displacement

Moment Frame Structure

Moment-Resisting Space Frame

Near-Source Factors

Nonbuilding Structure

Occupancy Category

Ordinary Moment-Resisting Space Frame

Orthogonal Effect

Overturning Moment

Plan Structural Irregularity

Rayleigh Method

Resisting Moment

Rigid Structure

Seismic Coefficient $C_a$

Seismic Coefficient $C_v$

Seismic Overstrength Factor, $\Omega_o$

Seismic Zone Factor

Separation Between Structures

Site Coefficient

Snow Load

Soft Story

Space Frame

Special Moment-Resisting Space Frame

Static Lateral Force Procedure

Steel Braced Frame

Steel Moment Frame

Story Drift

Story Drift Ratio

Structure System

UBC Design Criterion

UBC Design Ratio

Vertical Irregularity

Weak Story

Weight (Mass) Irregularity

Wind Load

1. The UBC bases its seismic zone factor on the effective peak ground acceleration that has a 10% chance of being exceeded in 50 years. What is this EPA intended to represent?

    A. the maximum probable earthquake
    B. the minimum considered (capable) earthquake
    C. the maximum predictable earthquake
    D. the minimum theoretical earthquake

2. In designing and detailing structures, the purpose of complying with UBC seismic requirements is to prevent which of the following in the event of an earthquake?

    A. structural damage
    B. architectural damage
    C. loss of life
    D. disruption of business

3. By what amount does the actual seismic force that can be developed in a structure exceed the design seismic force calculated using UBC principles?

    A. between 1.0 and 1.3
    B. between 1.3 and 1.6
    C. between 1.6 and 2.0
    D. between 2.0 and 2.8

4. UBC seismic provisions are intended as minimum requirements. The level of protection and safety can be magnified by increasing which of the following?

    A. the inherent overstrength and global ductility
    B. the design lateral force
    C. the redundancy
    D. all of the above

5. An engineering firm is designing a family dwelling in seismic zone 4. This area is subject to very strong winds. What loads should the firm consider in designing and detailing the structure?

    A. It should design for only wind load and comply with UBC seismic detailing requirements.
    B. It should design for only seismic load and comply with UBC seismic detailing requirements.
    C. It should design for the combination of both loads.
    D. It should design for the larger of the two loads and comply with UBC seismic detailing requirements.

6. Which of the following statements is not correct?

    A. Wind loads may govern over seismic loads.
    B. Seismic loads may govern over wind loads.
    C. Seismic loads are inertial in nature.
    D. Wind loads are inertial in nature.

7. A site is subject to liquefaction and earthslides and is located on an active major fault line. The UBC prevents construction on this site due to which of the following?

    A. liquefaction
    B. earthslides
    C. fault lines
    D. The UBC does not prevent construction on this site.

8. For concrete, when considering earthquake forces with the alternative basic load combinations, what percentage increase for all allowable stresses and soil bearing values does the UBC specify for the working stress design method?

    A. 25%
    B. 33%
    C. 50%
    D. 66%

9. Under what section of the UBC is *Deformation Compatibility* defined?

    A. Section 1633
    B. Section 1633.2
    C. Section 1633.2.4
    D. Section 1633.2, Item 4

10. Which of the following facilities must remain operational after an earthquake?

    A. police training centers
    B. dairies
    C. bakeries
    D. hospitals with surgery rooms

11. A small hospital equipped with a surgery room is located in a small town in California (seismic zone 4). Based on UBC requirements, how is this facility best identified?

    A. hazardous
    B. essential
    C. standard
    D. special

12. Per UBC requirements, what values of seismic coefficients $C_a$ and $C_v$, respectively, would be used when the soil profile type is unknown in seismic zone 3?

    A. 0.24 and 0.24
    B. 0.33 and 0.45
    C. 0.36 and 0.54
    D. 0.36 and 0.84

13. What does the $R$ factor account for?

    A. the resistance of the structural system
    B. the energy absorption capacity of the structural system
    C. the stiffness of the structural system
    D. all of the above

14. The value of $h_n$ for the following structures is

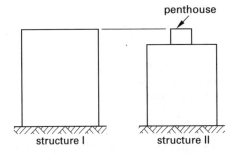

penthouse

structure I      structure II

    A. larger for structure I.
    B. larger for structure II.
    C. equal for structure I and structure II.
    D. $\Delta_M$ multiplied by the number of stories.

15. For steel eccentric braced frames, what is the value of $C_t$?

    A. 0.020
    B. 0.030
    C. 0.035
    D. 0.075

16. The total seismic dead load should include

    A. anything suspended inside the building from the roof.
    B. anything mounted on the upper half of the walls.
    C. permanent equipment mounted on the roof.
    D. all of the above.

17. For the base shear calculation of a building, an engineering firm needs to consider 40 lbf/ft$^2$ (1.92 kN/m$^2$)

of snow load with respect to the total weight, $W$. Which is the correct statement?

    A. Snow load can be excluded.
    B. Only 30 lbf/ft$^2$ (1.44 kN/m$^2$) of snow load shall be included.
    C. Snow load can be reduced 25%.
    D. Snow load may be reduced up to 75% with approval of the building official.

18. The weight, $W$, used in base shear calculation $(V = (C_v I/RT)W)$ excludes which of the following?

    A. 110 lbf/ft$^2$ (5.27 kN/m$^2$) of the floor live load in a warehouse
    B. 10 lbf/ft$^2$ (0.48 kN/m$^2$) of the floor partition load
    C. 40 lbf/ft$^2$ (1.92 kN/m$^2$) of the snow load
    D. none of the above

19. When designing a storage warehouse, what percent of live load must be included in the value of total weight, $W$?

    A. 0%
    B. 25%
    C. 75%
    D. 100%

20. For a proposed building in seismic zone 4, the study shows that the construction site is all hard rock-like material. The elastic fundamental period of vibration for the building is 1 sec (s). Using the following spectral shape, what will the corresponding value of the spectral acceleration be?

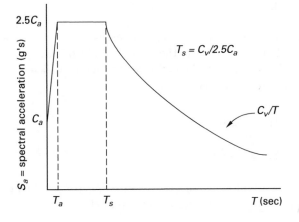

A. 0.32
B. 0.43
C. 0.54
D. 0.64

21. For determining the fundamental period, $T$, of a building with concrete shear walls by Method A, UBC Sec. 1630.2.2, Item 1 gives the following formula.

$$A_c = \sum A_e \left[ 0.2 + \left( \frac{D_e}{h_n} \right)^2 \right]$$

The value of $C_t$ for this structure may be taken as $0.1/\sqrt{A_c}$. (For SI, $0.0743/\sqrt{A_c}$ for $A_c$ in m$^2$.)

What is the maximum value of $D_e/h_n$?

A. 0.3
B. 0.9
C. 2.75
D. $\frac{3}{8} R_w$

22. For determining the fundamental period, $T$, of a building, the UBC gives the following formula.

$$T = 2\pi \sqrt{ \frac{ \sum\limits_{i=1}^{n} w_i \delta_i^2 }{ g \sum\limits_{i=1}^{n} f_i \delta_i } }$$

This formula represents

A. the period defined by Method A.
B. the period defined by Method B.
C. the static period.
D. the dynamic period.

23. The natural period of a building is 4 sec. At the roof level of the structure, the force, $F_t$, which is additional to the base shear, should be determined by which of the following formulas?

A. $F_t = 0$
B. $F_t = 0.07TV$
C. $F_t = C_t(h_n)^{3/4}$
D. $F_t = 0.25V$

24. If a building's natural period, $T$, is 0.6 sec, what is the concentrated force at the top of the structure?

A. $F_t = 0$
B. $F_t = 0.06TV$
C. $F_t = 0.07TV$
D. $F_t = 0.25V$

25. The maximum base shear can be calculated from which of the following formulas?

A. $V = \left( \dfrac{C_v I}{RT} \right) W$

B. $V = \left( \dfrac{2.5\, C_a I}{R} \right) W$

C. $V = 011 C_a I W$

D. $V = \left( \dfrac{0.8 Z N_v I}{R} \right) W$

26. Which of the following systems is without a complete vertical load-carrying space frame?

A. bearing wall systems
B. moment-resisting frames
C. building frame systems
D. dual systems

27. For a moment-resisting structural system, resistance to lateral load is provided by which of the following?

A. complete space frames
B. shear walls
C. braced frames
D. flexural action of members

28. Which of the following structural systems have the lowest ductility for structures three stories or less?

A. concrete shear walls
B. steel braced bearing walls
C. heavy timber braced bearing walls
D. wood structural panels bearing walls

29. Concrete is preferred over steel for designing a seven-story apartment complex in Los Angeles. For this building, what type of moment-resisting concrete frame should be selected?

A. special
B. intermediate
C. ordinary
D. any of the above

30. For a building in San Francisco, different types of structural systems are being analyzed. Since the height of the building is 200 ft (61 m), which of the following combinations are allowed by the UBC?

    I. dual systems
    II. special moment-resisting frames
    III. building frame systems

A. I only
B. I and II
C. I and III
D. I, II, and III

31. A structure consists of a special moment-resisting steel frame and shear walls. According to UBC requirements, what is the lowest percentage of base shear that the special moment-resisting steel frame must resist independently?

A. 0%
B. 25%
C. 50%
D. 75%

32. Consider an 80 ft (24.4 m) special moment-resisting steel frame. Applying Method A of UBC, Sec. 1630.2.2, what is the period of the building?

A. 0.40 sec
B. 0.53 sec
C. 0.80 sec
D. 0.94 sec

33. Three types of structures are shown. They are located at a site in seismic zone 4 and underlain by the same soil profile. If an earthquake occurs with a predominant period of 0.6 sec, which of the following buildings has the greatest possibility of being in resonance with the earthquake?

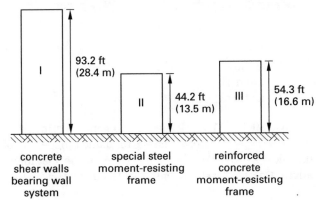

concrete shear walls bearing wall system    special steel moment-resisting frame    reinforced concrete moment-resisting frame

A. I only
B. I and III
C. II and III
D. I, II, and III

34. For a ten-story office building underlain with soil profile type $S_B$ in Los Angeles, $R = 5.6$, $W = 2000\ K$ (907 185 kg), $N_a = 1.0$, and $N_v = 1.0$. What is the minimum allowable design base shear for this structure?

A. 88.0 k  (390 kN)
B. 115.0 k  (510 kN)
C. 230.0 k  (1020 kN)
D. 393.0 k  (1750 kN)

35. For a structure with a seismic base shear coefficient $C_v I/RT = 0.14$, what is the effective spectral acceleration (expressed in percentage of gravity)?

A. 5%
B. 10%
C. 14%
D. 25%

36. A special moment-resisting steel structure is being designed with a height of 240 ft (73.2 m). The site for this building is underlain with very dense soil and soft rock with 2500 ft/sec (760 m/s) shear wave velocity in seismic zone 4. Based on the UBC requirement, the value of $C_v/T$ in designing base shear need not exceed which of the following?

A. $0.56 N_v$
B. $0.96 N_v/T$
C. $1.92/T$
D. $2.5 C_a$

37. A building must be designed that will not require any additional force at the roof level ($F_t = 0$). Assume the maximum possible natural period for this building. Use $R = 5.5$, soil profile type $A$, $Z = 0.4$, and near source factor $N_v = 1.2$. What is the design base shear using UBC Formula 30-4?

A. 0
B. $0.10 IW$
C. $1.0 IW$
D. $10 IW$

38. Considering UBC Table 16-P, for which of the following structures can the UBC Formula $F_p = 4.0 C_a I_p W_p$ not be used?

A.  penthouses
B.  water tanks
C.  braced parapet walls
D.  roof-mounted equipment

39. What are the minimum and maximum values of $R_p$ for the attachments of permanent equipment supported by a structure? Note that attachments include anchorages and required bracing.

    A.  1.0 and 3.0
    B.  1.0 and 1.5
    C.  1.5 and 3.0
    D.  1.5 and 4.0

40. For the design of the monument in the following illustration, what $R$ value should be used?

    A.  2.0
    B.  2.2
    C.  2.9
    D.  3.6

41. For special moment-resisting space frames, what are the values of $a_p$ and $R_p$, respectively, for attachment of the suspended ceilings? Note that attachments include anchorages and required bracing.

    A.  1.0 and 3.0
    B.  1.0 and 4.0
    C.  2.5 and 3.0
    D.  2.5 and 4.0

42. You are going to design a billboard next to a freeway. This billboard is a self-supporting structure. What value of the seismic force overstrength factor should be used?

    A.  2
    B.  3
    C.  4
    D.  5

43. Nonbuilding structures other than "rigid structures" and "tanks with supported bottoms" should be designed to resist design seismic forces not less than

    A.  $0.11C_aIW$
    B.  $0.56C_aIW$
    C.  $0.70C_aIW$
    D.  $3.0C_aW/R$

44. You are to design a rigid nonbuilding structure with a period $T = 0.04$ sec (0.04 s). Which UBC formula would you use for obtaining the lateral force?

    A.  $V = 0.7C_aIW$
    B.  $V = 0.11C_aIW$
    C.  $V = ZICW$
    D.  $V = C_vIW/RT$

45. The fundamental period of nonbuilding structures should be determined by which of the following UBC methods?

    A.  Method A
    B.  Method B
    C.  either method
    D.  neither method

46. Two water tanks are shown. One tank is well anchored on the top of a building. The second tank is on the ground and is self-supporting. Which tank is considered a nonbuilding structure?

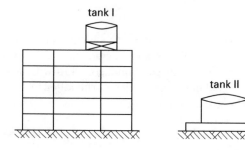

    A.  I only
    B.  II only
    C.  I and II
    D.  neither

47. An 80 ft (24.4 m) special moment-resisting steel frame with eight stories has a fundamental period of 0.7 sec (0.7 s). Per the UBC, what is the maximum allowed design level response displacement (total story drift)?

A. 0.30 in (8 mm)
B. 0.40 in (10 mm)
C. 0.48 in (12 mm)
D. 0.60 in (15 mm)

48. The calculated story drift of one level relative to the level above or below due to the design lateral force shall include

    I. torsional deflections.
    II. translational deflections.
    III. bending deflections.

A. I only
B. I and II
C. I and III
D. II and III

49. A structure in seismic zone 4 has an $R$ value of 8.5. What may be the maximum design height of this structure?

A. 65 ft (20 m)
B. 160 ft (49 m)
C. 240 ft (73 m)
D. no limit

50. A regular structure is proposed for a high-rise in seismic zone 3. A static analysis is performed to design this high-rise. What should the maximum design height be?

A. 65 ft (20 m)
B. 160 ft (49 m)
C. 240 ft (73 m)
D. no limit

51. An ordinary steel-braced frame structure will be designed for an insurance company in San Francisco. Based on UBC requirements, what should the maximum allowable height be for this building?

A. 65 ft (20 m)
B. 160 ft (49 m)
C. 240 ft (73 m)
D. no limit

52. A property owner desires to build an apartment building in San Francisco for a total occupancy load of 6000 persons. An irregular structure will be designed to meet the owner's need. If dynamic lateral force procedures are used, what should be the maximum allowable height?

A. 5 stories
B. 65 ft (20 m)
C. 240 ft (73 m)
D. no limit

53. An engineer computes the total base shear for a regular structure using the static force procedure, and Method B for the building period. If a dynamic analysis is performed using the default spectral response of UBC Fig. 16-3, what percentage reduction of the total base shear does the UBC allow?

A. 0%
B. 10%
C. 25%
D. 75%

54. For irregular structures, what percentage reduction of the total base shear (calculated by the static force procedure) does the UBC allow when dynamic analysis is performed?

A. 0%
B. 10%
C. 25%
D. 50%

55. An engineering firm is designing a hospital in seismic zone 4. The building has a vertical geometric irregularity. The height of the building is 65 ft (19.8 m). Which type of lateral force procedure should be used?

A. static
B. dynamic
C. either static or dynamic
D. neither

56. For the structure shown, $S_n$ represents the story strength of the $n^{\text{th}}$ floor. The relative story strengths of different floors are given as follows. Which floor should be identified as a weak story?

$$80\% \, S_2 < S_1 < 90\% \, S_2$$
$$80\% \, S_3 < S_2 < 85\% \, S_3$$
$$70\% \, S_4 < S_3 < 80\% \, S_4$$
$$S_4 = 80\% \, S_5$$

A.  1<sup>st</sup> floor
B.  2<sup>nd</sup> floor
C.  3<sup>rd</sup> floor
D.  4<sup>th</sup> floor

57. The horizontal dimensions of the lateral force-resisting systems for two structures are shown. Based on UBC Chap. 16, which building can be defined as a vertically irregular structure?

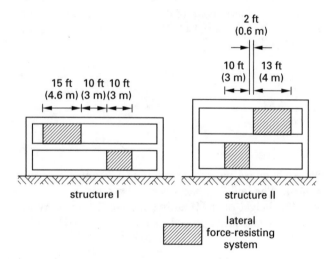

A.  I only
B.  II only
C.  I and II
D.  neither

58. A building is vertically irregular. For the fourth floor, the story strength is less than 80% of the story above. Which type of irregularity applies?

A.  soft story
B.  weak story
C.  mass story
D.  none of the above

59. Which of the following buildings has a soft story?

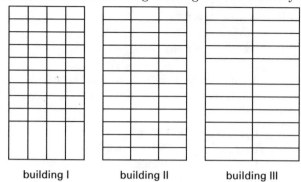

A.  I only
B.  I and II
C.  I and III
D.  II and III

60. In general, vertically irregular structures with more than two stories [or more than 30 ft (9.1 m) in height] are prohibited by the UBC when the percentage of one story's strength to the strength of the story above it is less than

A.  50%.
B.  65%.
C.  80%.
D.  85%.

61. A structure having a flexible upper portion (tower) supported on a rigid lower portion (platform) is shown. The tower and platform portions of the structure are classified as being regular. The periods shown are for each portion (the upper portion considered as a separate structure fixed at the base). The static lateral force procedure should be used. What is the maximum period for the entire structure?

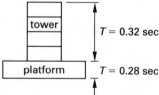

A.  0.28 sec
B.  0.32 sec
C.  0.35 sec
D.  0.50 sec

62. For the structures shown, the lower portions support the upper portions. The upper portions are flexible, while the lower portions are rigid. Each portion of

the structures is classified as being regular. Per UBC Sec. 1629, which of the structures can be designed by the static lateral force procedure, provided that the period of the entire structure is not greater than 1.1 times the period of the upper portion?

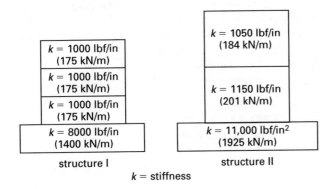

structure I

structure II

$k$ = stiffness

A. structure I only
B. structure II only
C. both structure I and structure II
D. neither

63. An 80 ft (24.4 m) structure in California has a concrete bearing wall system along one principal axis and a special moment-resisting steel frame along the orthogonal principal axis. What value of $R$ should be used?

A. 4.5
B. 5.5
C. 6.4
D. 8.5

64. In designing structures that support flexible nonstructural elements in seismic zones 3 and 4, when should interaction effects between the structure and the supported elements be considered?

A. combined weight exceeds 25% of the weight of the structure
B. combined weight does not exceed 25% of the weight of the structure
C. combined weight exceeds 75% of the weight of the structure
D. combined weight does not exceed 75% of the weight of the structure

65. Using allowable stress design loads, the UBC minimum resisting force for retaining walls against overturning is

A. half of the overturning moment.
B. $\frac{3}{4}$ of the overturning moment.
C. equal to the overturning moment.
D. 1.5 of the overturning moment.

66. Using allowable stress design loads, the UBC requires retaining walls to be designed to resist sliding by at least

A. 50% of the lateral force.
B. 85% of the lateral force.
C. 100% of the lateral force.
D. 150% of the lateral force.

67. Per UBC requirements, interior walls, permanent partitions, and temporary partitions that exceed 6 ft (1.8 m) in height should be designed to which of the following criteria?

I. to resist all loads to which they are subjected
II. to resist a minimum force of 5 lbf/ft² (0.24 kN/m²) applied perpendicular to the walls
III. to deflect a maximum of $L/360$ of the walls' span

A. I only
B. I and II
C. I and III
D. II and III

68. The structures shown are in seismic zone 4. During an earthquake, they will drift equally (i.e., the design level response displacements are the same, $\Delta_{SI} = \Delta_{SII} = \Delta_S$). There is no justification for rational analyses. Based on this information, what is the minimum required separation between these structures?

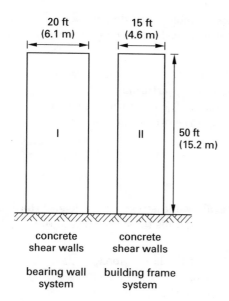

20 ft (6.1 m)  15 ft (4.6 m)

I

II

50 ft (15.2 m)

concrete shear walls

concrete shear walls

bearing wall system

building frame system

A. two times the displacement $\Delta_s$
B. three times the displacement $\Delta_s$
C. four times the displacement $\Delta_s$
D. five times the displacement $\Delta_s$

69. In San Francisco (seismic zone 4), a special moment-resisting steel frame structure with a height of 60 ft (18.3 m) is located next to a vacant space on the same property. An investor wants to build an 80 ft (24.4 m) concrete shear bearing wall structure on the lot, close to the first building. The fundamental periods of the existing and proposed buildings are 0.8 and 0.5 sec (s), respectively. Based on this information, what is the minimum separation between the two buildings required by the UBC? (There is no justification for rational analyses.) Assume each building has five equal stories.

*5 STORIES/BLDG*

A. 3 in (7 cm)
B. 23 in (60 cm)
C. 28 in (71 cm)
D. 35 in (89 cm)

70. Mechanical equipment is constructed on the top level (roof) of a building. The equipment is well attached to the structure. It weighs 30 k (13 562 kg). With respect to grade, the structure roof elevation is 80 ft (24.4 m), while the equipment attachment elevation is 86 ft (26.2 m). Using seismic zone 3 design criteria, determine the total design seismic force that this permanent nonstructural component and its attachment should resist. There is no analysis or empirical data available.

A. 0 lbf (0 N)
B. 7560 lbf (33 500 N)
C. 15,210 lbf (67 370 N)
D. 48,000 lbf (213 000 N)

71. Backup electrical equipment for a surgery room of a San Francisco hospital is installed on the roof with adequate support and anchorage. The closest distance between this site and the seismic source type A is 3.11 mi (5 km). The equipment weighs 4000 lbf (1814 kg). What is the lateral force on the equipment?

A. 3600 lbf (16 010 N)
B. 6230 lbf (27 710 N)
C. 7200 lbf (32 020 N)
D. 12,720 lbf (56 590 N)

72. For nonbuilding structures with special and standard occupancy categories in seismic zones 3 and 4, the intermediate moment-resisting frame (IMRF) may be used when

I. the structure is less than 50 ft (15.24 m) in height.
II. the structure is less than 160 ft (48.72 m) in height.
III. the value of $R$ used in reducing calculated member forces and moments does not exceed 2.8.

A. I only
B. II only
C. I and III
D. IMRF systems are prohibited in seismic zones 3 and 4.

73. A walkway tunnel connects two high-rise office buildings in San Francisco. What connections should be used between the walkway and the buildings?

A. fixed connections on both ends
B. hinges on both ends
C. sliding connections on both ends
D. sliding connections on one end and hinges on the other end

74. An engineer under supervision of a registered civil engineer can prepare which of the following?

    A. plans
    B. reports
    C. quantity calculations
    D. all of the above

75. Which of the following final documents must bear the seal or stamp of a registered civil engineer?

    I. plans
    II. specifications
    III. reports

    A. I only
    B. I and II
    C. I and III
    D. I, II, and III

76. A report characterizing the potential for liquefaction is needed for the site of an office building. Who provides the information?

    A. an architect
    B. a contractor
    C. a civil engineer
    D. a geotechnical engineer

77. For a single family home, who is responsible for signing the final soils report?

    A. a geotechnical engineer
    B. an architect
    C. a registered civil engineer
    D. a contractor

78. Who should sign the plans, specifications, and reports of an engineering firm?

    A. the owner of the firm
    B. the property owner
    C. a registered civil engineer
    D. a building inspector

79. For one-story residential buildings, which of the following registered professionals can design and sign plans?

    A. architects
    B. civil engineers
    C. structural engineers
    D. all of the above

80. Who should sign, stamp, and seal the structural design plans for a school in California?

    A. a registered civil engineer
    B. a registered structural engineer
    C. an experienced engineer
    D. a geotechnical engineer

81. For structural systems, the details of connections that resist seismic forces should be

    A. provided by the contractor.
    B. designed by the engineer.
    C. dictated by the building inspector.
    D. all of the above.

82. What does the California Hospital Act require?

    A. no construction of hospitals near faults
    B. fully functional hospitals after an earthquake
    C. using special moment-resisting frames for hospitals
    D. construction of hospitals with base isolation

83. The California legislature act that pertains to enforcing minimum standards for lateral force resistance in all structures is

    A. the Field Act.
    B. the Riley Act.
    C. the UBC Act.
    D. none of the above.

84. What California Administrative Code requires that hospitals be operational after an earthquake?

    A. Title 6
    B. Title 12
    C. Title 18
    D. Title 24

# CHAPTER 2
## SOLUTIONS

1. Answer A
The UBC's provisions are set as minimum requirements, and section 1627 defines the design basis ground motion (maximum probable) as that ground motion with a 10% chance of being exceeded in 50+ years, as determined by a site-specific hazard analysis or from a seismic hazard map.

2. Answer C
Based on the UBC [Sec. 101.2], the purpose of complying with UBC seismic requirements is to prevent loss of life in the event of an earthquake. Although the seismic codes affect the performance of structures and tend to prevent structural damage during an earthquake, this is not the main intent of the codes. According to UBC [Sec. 1626.1], the primary purpose of earthquake provisions is to avoid major structural failures in order to safeguard lives, not to limit damage or maintain function.

3. Answer D
In a linear elastic structure, the force in the structure reaches $R$ times the calculated design seismic force where $R$ is the global ductility factor of the structure and has a value between 2.2 and 8.5, depending on the structural system. However, because of the inelastic behavior of the actual structure, the natural period and the damping ratio increase and seismic energy is dissipated. This reduces the actual seismic force that can be developed to $\Omega_o$ times the design seismic force, where $\Omega_o$ is the force amplification factor that varies between 2.0 and 2.8, depending on the structure.

4. Answer D
The level of protection and safety applies to the seismic resistance of the structures, which can be increased by increasing the following.

- $R$ (the inherent overstrength and global ductility capacity of lateral-force resisting system)
- the design lateral force
- redundancy (i.e., two or more lateral load supporting systems)
- the quality of the construction materials
- the quality of the construction methods

5. Answer D
Per the UBC [Sec. 1626.3], between seismic and wind loads, the firm should design for the larger load and comply with the detailing provisions of the seismic codes.

6. Answer D
Per the UBC [Sec. 1626.3], the larger load, wind or seismic, governs the design. Wind loads are applied over the exterior surface of the structure, while the seismic loads are inertial in nature.

7. Answer D
This land is subject to liquefaction, earthslides, and major ground acceleration. The code does not prevent construction on this site, but as a minimum requirement, the UBC provisions should apply to the construction. To increase the level of protection, $R$, the design lateral force, redundancy, and the quality of the construction materials and construction methods should be increased. Other state and local codes may be applicable.

8. Answer B
With the alternative basic load combinations for concrete design, the UBC [Sec. 1612.3.2] allows for a $\frac{1}{3}$ increase when considering the earthquake force, either

acting alone or when combined with vertical loads. In the case of vertical loads acting alone, this increase is not permitted.

**9. Answer C**
*Deformation compatibility* is defined in the UBC [Sec. 1633.2.4].

**10. Answer D**
Per UBC Table 16-K, only those facilities that are categorized as occupancy category 1 (Essential Facilities) need to remain operational after an earthquake. Hospitals with surgery and emergency rooms are essential facilities and should remain open.

**11. Answer B**
Per UBC Table 16-K, for occupancy category 1, hospitals having surgery and emergency treatment areas are considered essential facilities.

**12. Answer C**
Soil profile types $S_A$, $S_B$, $S_C$, $S_D$, $S_E$, and $S_F$ are defined in UBC Table 16-J and soil profile type $S_F$ is defined as soils requiring site-specific evaluation. The UBC [Sec. 1629.3.1] identifies soils that can be classified as the soil profile type $S_F$.

However, according to UBC [Sec. 1636.2], soil profile type $S_D$ should be used for locations with insufficient geotechnical details to determine the soil profile type.

For the soil profile type $S_D$ in seismic zone 3, based on UBC Table 16-Q, the seismic coefficient $C_a$ is 0.36 and based on UBC Table 16-R, the seismic coefficient $C_v$ is 0.54.

**13. Answer B**
The response modification factor $R$ accounts for the energy absorption capacity. Per the UBC [Sec. 1628], it is a numerical coefficient representative of the inherent overstrength and global ductility capacity of lateral-force resisting systems. $R$ values are tabulated in UBC Table 16-N, for various types of building structural systems, and in Table 16-P for nonbuilding structures. As the value of $R$ increases, the ductility of the structure increases.

**14. Answer A**
Per the UBC [Sec. 1628], $h_n$ is defined as the height in feet (m) above the base to the $n^{th}$ level. The $n^{th}$ level is the uppermost level in the main portion of the building. The penthouse is not considered part of the main portion of the structure, so $h_n$ is larger for structure I.

**15. Answer B**
The UBC [Sec. 1630.2.2, Item 1] gives 0.030 (0.0731) for the value of $C_t$ for eccentrically braced steel frames.

**16. Answer D**
Based on the UBC [Sec. 1630.1.1], the total seismic dead load of the structure includes the following.

- all elements of the structure that are permanent, such as roofs, walls, floors, and equipment

- all elements suspended inside the building from the roof or ceiling, or mounted on the upper half of the walls

- applicable portions of other loads, such as floor live loads, partition loads, and snow loads

**17. Answer D**
Where the snow load is greater than 30 lbf/ft$^2$ (1.44 kN/m$^2$), the UBC [Sec. 1630.1.1, Item 3] specifies that the design snow load should be included in the total seismic load. However, with the approval of building officials, the design snow load may be reduced up to 75%.

**18. Answer D**
Per the UBC [Secs. 1630.1.1, Items 1, 2, and 3], for the base shear calculation, the weight, $W$, should include a minimum of 25% of the floor live load in a warehouse, a minimum of 10 lbf/ft$^2$ (0.48 kN/m$^2$) for floor partitions, and the snow load if it is greater than 30 lbf/ft$^2$ (1.44 kN/m$^2$).

**19. Answer B**
Per the UBC [Sec. 1630.1.1, Item 1], in warehouse occupancies, a minimum of 25% of the floor live load must be accounted for in the value of the total weight, $W$.

**20. Answer D**
Based on Table 16-J, the soil profile type for "hard rock" is $S_A$. From UBC Table 16-Q, the seismic response coefficient $C_a$ for seismic zone 4 and soil profile type $S_A$ is 0.32 $N_a$. Similarly, the value of seismic response coefficient $C_v$ is obtained from UBC Table 16-R as 0.32 $N_a$.

The near-source factor used in the determination of $C_a$ in seismic zone 4 related to both the proximity of the building to known faults with magnitudes and slip rates as given in UBC Table 16-S and 16-U. Since this information is unknown here, from UBC Table 16-S, the maximum near-source factor of 1.5 should be used to find the maximum value of $C_v$. Therefore,

$$C_a = (0.32)(1.5)$$
$$= 0.48$$

Similarly, from Table 16-T, $N_v = 2.0$ and $C_v = (0.32) \times (2.0) = 0.64$.

From the given spectral shape,

$$T_s = \frac{C_v}{2.5C_a}$$
$$= \frac{0.64}{(2.5)(0.48)}$$
$$= 0.533 \text{ sec}$$

0.533 sec is less than the elastic fundamental period of vibration for the building (1 sec). Hence,

$$S_a = \frac{C_v}{T}$$
$$= \frac{0.64}{1.0 \text{ sec}}$$
$$= 0.64$$

Refer to the UBC [Secs. 1627 and 1631.2] and UBC Figure 16-3 for a more in-depth understanding of the design response spectrum.

**21. Answer B**
Method A of the UBC [Sec. 1630.2.2, Item 1] details that the value of $D_e/h_n$ should not exceed 0.9 when used in UBC Formula 30-9.

**22. Answer B**
Method B, as defined in the UBC [Sec. 1630.2.2, Item 2] uses UBC Formula 30-10 to calculate the fundamental period, $T$. The terms *static period* and *dynamic period* have no meaning.

**23. Answer D**
Per the UBC [Sec. 1630.5], if the value of $T$ is 0.7 sec or less, the additional force at the top, $F_t$, is equal to zero. Otherwise, use UBC Formula 30-14.

$$F_t = 0.07TV$$

For this particular structure, the natural period of 4 sec is greater than 0.7 sec, so Formula 30-14 is applicable. $F_t$ need not exceed $0.25V$, so perform the following multiplication.

$$F_t = 0.07TV$$
$$= (0.07)(4 \text{ sec})V$$
$$= 0.28V$$

Since $0.28V$ is greater than the maximum value of $0.25V$, use $F_t = 0.25V$.

**24. Answer A**
Based on the UBC [Sec. 1630.5], use UBC Formula 30-14 for determining the force, at the top, $F_t$. Formula 30-14 is $F_t = 0.07TV$. This formula should be used where $T$ is greater than 0.7 sec. If $T$ is equal to or less than 0.7 sec, $F_t$ can be considered to equal zero.

**25. Answer B**
For base shear calculations in a given direction, UBC Formula 30-4, $V = (C_vI/RT)W$, should be used. Per the UBC [Sec. 1630.2.1], the maximum base shear can be calculated from $V = (2.5C_aI/R)W$. $C_a$ and $C_v$ are seismic coefficients and their values can be obtained from UBC Tables 16-Q and 16-R, respectively. The maximum value of $C_v/T$ is $2.5C_a$.

**26. Answer A**
Per the UBC [Sec. 1629.6.2], *bearing wall systems* are defined as those structural systems that are without a complete vertical load-carrying frame. Bearing walls give support for all or most of the gravity loads, and lateral loads are resisted by shear walls or braced frames.

**27. Answer D**
These systems have an essentially complete frame that gives support to both vertical and lateral loads. According to the UBC [Sec. 1629.6.4], resistance to lateral load is provided by flexural actions of members.

**28. Answer C**
The response modification factor $R$ has been assigned to reflect the inherent overstrength and global ductility capacity of different lateral-force resisting systems. As the value of $R$ decreases, the ductility of the structure decreases. For $R$ values, refer to the UBC, Table 16-N. Using this table, the values of $R$ for choices (A), (B), (C), and (D) are 4.5, 4.4, 2.8, and 5.5, respectively.

**29. Answer A**
There are three types of concrete moment-resisting frames: special, intermediate, and ordinary. These concrete frames are required by design to be part of the lateral-force resisting system. The UBC permits these systems to be constructed only in certain seismic zones. For Los Angeles (seismic zone 4), per UBC Table 16-N, ftns. 5 and 8, and [Sec. 1633.2.7, Item 1], the special moment-resisting concrete frame is the only choice available.

**30. Answer B**

The UBC [Sec. 1630.4.3] allows only combinations of dual systems and special moment-resisting frames to be used to resist seismic forces in seismic zones 3 and 4, and for structures exceeding 160 ft (48.8 m) in height.

**31. Answer B**

Since this structure consists of a combination of a special moment-resisting steel frame and shear walls, the UBC [Sec. 1629.6.5] defines it as a *dual system*. For dual systems, the UBC [Sec. 1629.6.5, Item 2] specifies that the moment-resisting frames shall be designed to independently resist at least 25 percent of the design base shear.

**32. Answer D**

*SI solution*

From the UBC [Sec. 1630.2.2, Item 1], the value of $C_t$ given for steel moment-resisting frames is 0.0853. Use UBC Formula 30-8 (Method A) to find the building period.

$$T = C_t(h_n)^{3/4}$$
$$= (0.0853)(24.4 \text{ m})^{3/4}$$
$$= 0.94 \text{ s}$$

Note that the value of $h_n$ is defined as the height of the building in meters.

*Customary U.S. solution*

From the UBC [Sec. 1630.2.2, Item 1], the value of $C_t$ given for steel moment-resisting frames is 0.035. Use UBC Formula 30-8 (Method A) to find the building period.

$$T = C_t(h_n)^{3/4}$$
$$= (0.035)(80 \text{ ft})^{3/4}$$
$$= 0.94 \text{ sec}$$

Note that the value of $h_n$ is defined as the height of the building in feet.

**33. Answer D**

*SI solution*

Resonance occurs when the building period coincides with the earthquake period. Find the period of each structure and compare it with the earthquake period of 0.6 s. Use UBC Formula 30-8, $T = C_t(h_n)^{3/4}$. From the UBC [Sec. 1630.2.2, Item 1], choose the appropriate value of $C_t$ for each structure. For structure I, $C_t = 0.0488$; for structure II, $C_t = 0.0853$; and for structure III, $C_t = 0.0731$.

$$T_1 = (0.0488)(28.4 \text{ m})^{3/4} = 0.6 \text{ s}$$
$$T_2 = (0.0853)(13.5 \text{ m})^{3/4} = 0.6 \text{ s}$$
$$T_3 = (0.0731)(16.6 \text{ m})^{3/4} = 0.6 \text{ s}$$

All of the buildings have an equal fundamental period of vibration, and they are all located on the same soil profile; therefore, all three buildings have the same possibility of being in resonance with the earthquake.

*Customary U.S. solution*

Resonance occurs when the building period coincides with the earthquake period. Find the period of each structure and compare it with the earthquake period of 0.6 sec. Use UBC Formula 30-8, $T = C_t(h_n)^{3/4}$. From the UBC [Sec. 1630.2.2, Item 1], choose the appropriate value of $C_t$ for each structure. For structure I, $C_t = 0.020$; for structure II, $C_t = 0.035$; and for structure III, $C_t = 0.030$.

$$T_1 = (0.020)(93.2 \text{ ft})^{3/4} = 0.6 \text{ sec}$$
$$T_2 = (0.035)(44.2 \text{ ft})^{3/4} = 0.6 \text{ sec}$$
$$T_3 = (0.030)(54.3 \text{ ft})^{3/4} = 0.6 \text{ sec}$$

All of the buildings have an equal fundamental period of vibration, and they are all located on the same soil profile; therefore, all three buildings have the same possibility of being in resonance with the earthquake.

**34. Answer B**

*SI solution*

Per the UBC [Sec. 1630.2.1], the minimum design base shear in a given direction can be obtained from UBC Formula 30-6, $V = 0.11C_a I W$. In addition, for Los Angeles, located in seismic zone 4, the computed base shear should not be less than the base shear obtained from UBC Formula 30-7, $V = (0.8ZN_v I/R)W$.

From UBC Table 16-Q, the seismic coefficient $C_a$ is 0.40 $N_a$ for the soil profile type $S_B$ in seismic zone 4.

Use UBC Formula 30-6. From the values given in the problem statement,

$$V = (0.11)(0.40)(1.0)(907\,185 \text{ kg})\left(9.81 \frac{\text{m}}{\text{s}^2}\right)$$
$$= 391.5 \text{ kN}$$

Check UBC Formula 30-7. From the values given in the problem statement,

$$V = \left( \frac{(0.8)(0.4)(1.0)(1.0)}{5.6} \right) (907\,185 \text{ kg}) \left( 9.81\ \frac{\text{m}}{\text{s}^2} \right)$$

$$= 508.5 \text{ kN}$$

Since 391.5 kN is less than 508.5 kN, $V_{\min} = 508.5$ kN should be used.

*Customary U.S. solution*

Per the UBC [Sec. 1630.2.1], the minimum design base shear in a given direction can be obtained from UBC Formula 30-6, $V = 0.11\,C_a I W$. In addition, for Los Angeles, located in seismic zone 4, the computed base shear should not be less than the base shear obtained from UBC Formula 30-7, $V = (0.8ZN_r I/R)W$. Check SI solution.

From UBC Table 16-Q, the seismic coefficient $C_a$ is 0.40 $N_a$ for the soil profile type $S_\text{B}$ in seismic zone 4.

Use UBC Formula 30-6. From the values given in the problem statement,

$$V = (0.11)(0.40)(1.0)(1.0)(2000 \text{ k})$$

$$= 88 \text{ k}$$

Check UBC Formula 30-7. From the values given in the problem statement,

$$V = \left( \frac{(0.8)(0.4)(1.0)(1.0)}{5.6} \right) (2000 \text{ k})$$

$$= 114.30 \text{ k}$$

Since 88 k is less than 114.30 k, $V_{\min} = 114.30$ k should be used.

### 35. Answer C

*SI solution*

$$F = ma = \left( \frac{W}{g} \right) a$$

$$a = \left( \frac{F}{W} \right) g \qquad \text{[Eq. 1]}$$

$$V = \left( \frac{C_v I}{RT} \right) W = 0.14\ W$$

$$F = V$$

Therefore,

$$F = 0.14W \qquad \text{[Eq. 2]}$$

Substitute Eq. 2 into Eq. 1 and solve.

$$a = \frac{0.14Wg}{W} = 0.14g$$

The effective spectral acceleration in percentage of gravity is 0.14 gravities.

*Customary U.S. solution*

$$F = \left( \frac{m}{g_c} \right) a = \left( \frac{W}{g} \right) a$$

$$a = \left( \frac{F}{W} \right) g \qquad \text{[Eq. 1]}$$

$$V = \left( \frac{C_v I}{RT} \right) W = 0.14W$$

$$F = V$$

Therefore,

$$F = 0.14W \qquad \text{[Eq. 2]}$$

Substitute Eq. 2 into Eq. 1 and solve.

$$a = \left( \frac{0.14W}{W} \right) g = 0.14g$$

The effective spectral acceleration in percentage of gravity is 0.14 gravities.

### 36. Answer D

From UBC Table 16-J, the soil profile type is $S_\text{C}$ for dense soil and soft rock with the given shear wave velocity. From UBC Table 16-R, the value of seismic response coefficient $C_v$ is 0.56 $N_v$ for the soil profile type $S_\text{C}$, and the site is located in seismic zone 4. From UBC Table 16-T, the maximum value of near-source factor $N_a$ that can be used is 2.0. Thus, the maximum value of the seismic response coefficient $C_v$ is

$$C_v = (0.56)(2)$$

$$= 1.12$$

$$\frac{C_v}{T} = \frac{1.12}{T}$$

However, the UBC [Sec. 1630.2.1] specifies that the total design base shear in a given direction should be determined from the UBC Formula 30-4,

$$V = \left( \frac{C_v I}{RT} \right) W$$

Further, on this same section UBC clarifies that the calculated design base shear need not exceed UBC Formula 30-5,

$$V = \left( \frac{2.5C_a I}{R} \right) W$$

Therefore,

$$\left(\frac{C_v I}{RT}\right) W = \left(\frac{2.5 C_a I}{R}\right) W$$

$$\frac{C_v}{T} = 2.5 C_a$$

**37. Answer B**

Note that $C_v = 0.32 N_r$ because (1) $Z = 0.4$ and (2) the soil profile type is $S_A$.

From the UBC [Sec. 1630.5], the maximum value of $T$ for $F_t$ to be zero is 0.7 sec. From UBC Table 16-R, the seismic coefficient $C_v$ for $Z$ is 0.4, and soil profile type $S_A = 0.32 N_v$.

$$C_v = 0.32 N_v$$
$$= (0.32)(1.2)$$
$$= 0.384$$

Use UBC Formula 30-4,

$$V = \left(\frac{C_v I}{RT}\right) W$$
$$= \left(\frac{0.384 I}{(5.5)(0.7)}\right) W$$
$$= 0.10 I W$$

**38. Answer B**

UBC Table 16-P identifies nonbuilding structures, including water tanks. Per the UBC [Sec. 1634.1.1], nonbuilding structures include all self-supporting structures other than buildings that carry gravity loads and resist the effects of lateral forces. Penthouses, braced parapet walls, and roof-mounted equipment are considered elements of structures, nonstructural components and equipment, and their attachments. For the latter, UBC Formula 32-1, $F_p = 4.0 C_a I_p W_p$, determines the total design lateral seismic force. The value of $C_a$ should be based on UBC Table 16-Q for the soil profile type. The value of $I_p$ can be obtained from Table 16-K.

**39. Answer C**

The attachments for permanent equipment supported by a structure should be designed to resist the total design seismic forces outlined in the UBC [Sec. 1632.2].

Per the UBC [Sec. 1632.2], $R_p$ is the Component Response Modification Factor. A value for $R_p$ should be taken from UBC Table 16-O. This table refers to $R_p$ as the horizontal force factor and lists values varying from 1.5 to 4.0. For attachments including anchorages and required bracing, the maximum value is 3.0.

Based on the UBC [Sec. 1632.2], $R_p$ value for anchorages should equal 1.5 for shallow expansion anchor bolts, shallow chemical anchors, or shallow cast-in-place anchors. $R_p$ should equal 1.0 for anchorages that are constructed of nonductile materials or by use of adhesive.

**40. Answer B**

A monument is considered a nonbuilding structure. For amusement structures and monuments, the $R$ value is 2.2, as obtained from UBC Table 16-P.

**41. Answer A**

UBC Table 16-O gives $a_p$ and $R_p$ values for elements of structure, nonstructural components, and equipment. Anchorage for suspended ceilings has an $a_p$ value of 1.0 and an $R_p$ value of 3.0.

**42. Answer A**

Signs and billboards are known as nonbuilding structures. Per UBC Table 16-P, the seismic force overstrength factor, $\Omega_o$, is 2.0.

**43. Answer B**

Nonbuilding structures include all self-supporting structures other than buildings that carry gravity loads and resist seismic forces.

Provisions for "rigid structures" and "tanks with supported bottoms" are given in the UBC [Secs. 1634.3 and 1634.4, respectively].

Based on the UBC [Sec. 1634.5], other nonbuilding structures excluding rigid structures and tanks with supported bottoms should be designed to resist design seismic forces not less than those forces determined from UBC Formula 34-2, $V = 0.56 C_a I W$. Note that for seismic zone 4, the total design base shear should also not be less than UBC Formula 34-3, $V = (1.6 Z N_v I / R) W$.

**44. Answer A**

For nonbuilding structures, the UBC provisions of Sec. 1634 should be used. Rigid structures are defined as those with a period, $T$, less than 0.06 sec. In the lateral force design, UBC [Sec. 1634.3] Formula 34-1, $V = 0.7 C_a I W$, should be used. The value of $C_a$ is based on UBC Table 16-Q for the soil profile type.

**45. Answer B**

Based on the UBC [Sec. 1634.1.4], the fundamental period of nonbuilding structures should be calculated by rational methods such as Method B [Sec. 1630.2.2, Item 2].

**46. Answer B**
The UBC [Sec. 1634.1.1] defines *nonbuilding structures* as self-supporting structures (other than buildings) that carry gravity loads and resist lateral forces. Therefore, the first tank is supported by the structure, while the second tank is considered a nonbuilding structure.

**47. Answer B**

*SI solution*

$\Delta_S$, *design level response displacement*, is the total drift or total story drift that occurs when the structure is subjected to the design seismic forces.

Based on the UBC [Sec. 1630.10.1], story drift should be computed using the maximum inelastic response displacement ($\Delta_M$). $\Delta_M$ can be obtained from UBC Formula 30-17, $\Delta_M = 0.7R\Delta_S$.

Per the UBC [Sec. 1630.10.2], the calculated story drift using $\Delta_M$ should not exceed 0.020 times the story height for structures having a fundamental period of 0.7 s or greater.

$$\Delta_M = 0.020h$$
$$= (0.020)\left(\frac{24.4 \text{ m}}{8 \text{ stories}}\right)\left(1000 \text{ } \frac{\text{mm}}{\text{m}}\right)$$
$$= 61 \text{ mm}$$

From UBC Table 16-N, for special moment-resisting steel frame, $R = 8.5$.

To compute the maximum allowable design level response displacement, use UBC Formula 30-17.

$$\Delta_M = 0.7R\Delta_S$$
$$\Delta_S = \frac{\Delta_M}{0.7R}$$
$$= \frac{61 \text{ mm}}{(0.7)(8.5)}$$
$$= 10 \text{ mm}$$

*Customary U.S. solution*

$\Delta_S$, *design level response displacement*, is the total drift or total story drift that occurs when the structure is subjected to the design seismic forces.

Based on the UBC [Sec. 1630.10.1], story drift should be computed using the maximum inelastic response displacement ($\Delta_M$). $\Delta_M$ can be obtained from UBC Formula 30-17, $\Delta_M = 0.7R\Delta_S$.

Per the UBC [Sec. 1630.10.2], the calculated story drift using $\Delta_M$ should not exceed 0.020 times the story height for structures having a fundamental period of 0.7 sec or greater.

$$\Delta_M = 0.020h$$
$$= (0.020)\left(\frac{80 \text{ ft}}{8 \text{ stories}}\right)\left(12 \text{ } \frac{\text{in}}{\text{ft}}\right)$$
$$= 2.4 \text{ in}$$

From UBC Table 16-N, for special moment-resisting steel frame, $R = 8.5$.

To compute the maximum allowable design level response displacement, use UBC Formula 30-17.

$$\Delta_M = 0.7R\Delta_S$$
$$\Delta_S = \frac{\Delta_M}{0.7R}$$
$$= \frac{2.4 \text{ in}}{(0.7)(8.5)}$$
$$= 0.4 \text{ in}$$

**48. Answer B**
The *story drift* is defined as the lateral displacement of one level relative to the level above or below. For computing story drifts, the maximum inelastic response displacement, $\Delta_M$, should be used, $\Delta_M = 0.7R\Delta_S$. $\Delta_S$, the resulting deformation, should be determined at all critical locations in the structure according to the UBC [Sec. 1630.9.1]. The calculated drift, per the UBC [Sec. 1630.9.1], should include torsional and translational deflections.

**49. Answer D**
The structural systems and $R$ factors are listed in UBC Table 16-N. Based on the information in this table, an $R$ of 8.5 applies to special moment-resisting frames and some dual systems with special moment-resisting frames. In all seismic zones, there is no height limit applicable to these systems.

**50. Answer C**
For regular structures, when the static lateral force procedure is used, the maximum design height of the structures (per the UBC [Sec. 1629.8.3, Item 2]) should be 240 ft (73.2 m).

**51. Answer B**
Ordinary steel and concrete braced frames are classified as building frame systems. Based on UBC Table 16-N, a 160 ft (48.8 m) height limit is applicable to those systems with steel in all seismic zones. However, concrete ordinary frame systems are prohibited in seismic zones 3 and 4.

**52. Answer D**
When the dynamic lateral force procedure is used, the UBC [Sec. 1629.8.4] imposes no height restriction for irregular structures in any seismic zone.

**53. Answer B**
According to the UBC [Sec. 1631.5.4, Items 1 and 2], the base shear for regular structures, $V_{dynamic}$, computed by dynamic analyses procedures, should be as follows.

- greater than or equal to 80% $V_{static}$ (where the ground motion representation complies with the UBC [Sec. 1631.2, Item 2])
- greater than or equal to 90% $V_{static}$ (where the ground motion representation complies with the UBC [Sec. 1631.2, Item 1])

**54. Answer A**
According to the UBC [Sec. 1631.5.4, Item 3], the base shear for irregular structures, $V_{dynamic}$, computed by dynamic analyses procedures should be greater than or equal to $V_{static}$ (where $V_{static}$ is determined in accordance with the UBC [Sec. 1630.2]).

**55. Answer C**
For irregular structures with a vertical geometric irregularity in seismic zone 4,

- static force procedures may be used per the UBC [Sec. 1629.8.3, Item 3] if the structure is not more than five stories or 65 ft (19.8 m) in height.
- dynamic analyses procedures may be used per the UBC [Sec. 1629.8.4, Item 2] without a height limit.

**56. Answer C**
Per the UBC [Sec. 1629.9.1 and Table 16-L], vertical structural irregularities of the weak story type exist where the story strength is less than 80% of that in the story above, or $S_n < 80\% \ S_{n+1}$. Checking for each story shows that $S_3 < 80\% \ S_4$. Therefore, the third floor is a weak story.

**57. Answer C**

*SI solution*

UBC Table 16-L defines an "in-plane discontinuity in vertical lateral force-resisting element," where the offset of the elements is greater than the length of the elements.

Assume that $L_1$ is the length of the lateral force-resisting system in the first story, and $L_2$ is the length of the element in the second story. If the offset length between stories is greater than the element lengths, vertical structural irregularities exist.

For building I, the offset length is 3 m + 3 m = 6 m, and the element lengths are $L_1 = 3.0$ m and $L_2 = 4.6$ m. Since 6.0 m > 3.0 m, an irregularity exists. Also, since 6.0 m > 4.6 m, an irregularity exists.

For building II, offset length is 3 m + 0.6 m = 3.6 m, and the element lengths are $L_1 = 3.0$ m and $L_2 = 4.0$ m. Since 3.6 m > 3.0 m), an irregularity exists. But, 3.6 m < 4.6 m, which does not meet the criteria. However, the previous comparison with $L_1$ is sufficient to classify the structure as vertically irregular.

Therefore, both buildings are irregular structures. Structure I also qualifies as vertically irregular under Category C, Vertical Geometric Irregularity. Vertical geometric irregularity exists when the horizontal dimension of the lateral force-resisting system in any story is more than 130% of that in an adjacent story.

For building I, the element lengths are $L_1 = 3.0$ m and $L_2 = 4.6$ m.

$$\frac{L_2}{L_1} = \frac{4.6 \text{ m}}{3 \text{ m}} = 1.5$$

The horizontal dimension of the element in the second story is 150% of that in the first story, so the structure is vertically irregular.

For building II, the element lengths are $L_1 = 3.0$ m and $L_2 = 4.0$ m.

$$\frac{L_2}{L_1} = \frac{4 \text{ m}}{3 \text{ m}} = 1.3$$

The horizontal dimension of the element in the second story is 130% of that in the first story, which is not great enough to classify the structure as vertically irregular under this category.

*Customary U.S. solution*

UBC Table 16-L defines an "in-plane discontinuity in vertical lateral force-resisting element," where the offset of the elements is greater than the length of the elements.

Assume that $L_1$ is the length of the lateral force-resisting system in the first story, and $L_2$ is the length of the element in the second story. If the offset length between stories is greater than the element lengths, vertical structural irregularities exist.

For building I, the offset length is 10 ft + 10 ft = 20 ft, and the element lengths are $L_1 = 10$ ft and $L_2 = 15$ ft. Since 20 ft > 10 ft, an irregularity exists. Also, since 20 ft > 15 ft, an irregularity exists.

For building II, offset length is 10 ft + 2 ft = 12 ft, and the element lengths are $L_1 = 10$ ft and $L_2 = 13$ ft. Since 12 ft > 10 ft, an irregularity exists. But 12 ft < 13 ft, which does not meet the criteria. However, the previous comparison with $L_1$ is sufficient to classify the structure as vertically irregular.

Therefore, both buildings are irregular structures. Structure I also qualifies as vertically irregular under Category C, Vertical Geometric Irregularity. Vertical geometric irregularity exists when the horizontal dimension of the lateral force-resisting system in any story is more than 130% of that in an adjacent story.

For building I, the element lengths are $L_1 = 10$ ft and $L_2 = 15$ ft.
$$\frac{L_2}{L_1} = \frac{15\,\text{ft}}{10\,\text{ft}} = 1.5$$

The horizontal dimension of the element in the second story is 150% of that in the first story, so the structure is vertically irregular.

For building II, the element lengths are $L_1 = 10$ ft and $L_2 = 13$ ft.
$$\frac{L_2}{L_1} = \frac{13\,\text{ft}}{10\,\text{ft}} = 1.3$$

The horizontal dimension of the element in the second story is 130% of that in the first story, which is not great enough to classify the structure as vertically irregular under this category.

## 58. Answer B

UBC Table 16-L categorizes the irregularity types, gives their definitions, and provides their appropriate UBC reference sections. Based on this table and Sec. 1629.9.1, when the story strength is less than 80% that of the story above, the vertically irregular structure is identified as Type 5, Discontinuity in Capacity—Weak Story.

## 59. Answer C

The soft story, per UBC Table 16-L, is a type of vertical structural irregularity where the story stiffness is either less than 70% of that in the story above or less than 80% of the average stiffness of the three stories above. By inspection, the first story of building I and the middle story of building III are considered soft stories.

## 60. Answer B

Based on UBC Table 16-L, vertically irregular structures defined as Type 5, Discontinuity in Capacity—Weak Story, are those in which the story strength ratio of one story to the story above is less than 80%. Per the UBC [Sec. 1629.9.1], Type 5 structures with more than two stories or more than 30 ft (9.14 m) in height may not have a story strength ratio of less than 65%.

## 61. Answer C

*SI solution*

Per the UBC [Sec. 1629.8.3, Item 4], the static lateral force procedure of Sec. 1630 is appropriate where $T_{\text{entire structure}} \leq 1.1 T_{\text{upper portion}}$. Solve for the maximum period of the entire structure.

$$T_{\text{maximum}} \leq (1.1)(0.32\,\text{s}) = 0.35\,\text{s}$$

*Customary U.S. solution*

Per the UBC [Sec. 1629.8.3, Item 4], the static lateral force procedure of Sec. 1630 is appropriate where $T_{\text{entire structure}} \leq 1.1 T_{\text{upper portion}}$. Solve for the maximum period of the entire structure.

$$T_{\text{maximum}} \leq (1.1)(0.32\,\text{sec}) = 0.35\,\text{sec}$$

## 62. Answer B

*SI solution*

Per the UBC [Sec. 1629.8.3, Item 4], the static lateral force procedure of Sec. 1630 is permitted where the average story stiffness of the lower portion is at least 10 times the average story stiffness of the upper portion.

$$K_{\text{average,rigid lower portion}}$$
$$\geq 10 K_{\text{average,flexible upper portion}}$$

Check above criterion for both structures.

Structure I:

$$K_{\text{average,rigid lower portion}} = 1400 \text{ kN/m}$$

$$K_{\text{average,flexible upper portion}} = \frac{(3)\left(175 \; \dfrac{\text{kN}}{\text{m}}\right)}{3}$$
$$= 175 \text{ kN/m}$$

Since 1400 kN/m is not greater than or equal to 10 times 175 kN/m, the criterion is not met and the static lateral force procedure is not permitted.

Structure II:

$$K_{\text{average,rigid lower portion}} = 1925 \text{ kN/m}$$

$$K_{\text{average,flexible upper portion}} = \frac{184 \; \dfrac{\text{kN}}{\text{m}} + 201 \; \dfrac{\text{kN}}{\text{m}}}{2}$$
$$= 192.5 \text{ kN/m}$$

Since 1925 kN/m is equal to 10 times 192.5 kN/m, the criterion is met and the static lateral force procedure is permitted.

*Customary U.S. solution*

Per the UBC [Sec. 1629.8.3, Item 4], the static lateral force procedure of Sec. 1630 is permitted where the average story stiffness of the lower portion is at least 10 times the average story stiffness of the upper portion.

$$K_{\text{average,rigid lower portion}}$$
$$\geq 10 K_{\text{average,flexible upper portion}}$$

Check above criterion for both structures.

Structure I:

$$K_{\text{average,rigid lower portion}} = 8000 \text{ lbf/in}$$

$$K_{\text{average,flexible upper portion}} = \frac{(3)\left(1000 \; \dfrac{\text{lbf}}{\text{in}}\right)}{3}$$
$$= 1000 \text{ lbf/in}$$

Since 8000 lbf/in is not greater than or equal to 10 times 1000 lbf/in, the criterion is not met and the static lateral force procedure is not permitted.

Structure II:

$$K_{\text{average,rigid lower portion}} = 11,000 \text{ lbf/in}$$

$$K_{\text{average,flexible upper portion}} = \frac{1050 \; \dfrac{\text{lbf}}{\text{in}} + 1150 \; \dfrac{\text{lbf}}{\text{in}^2}}{2}$$
$$= 1100 \text{ lbf/in}$$

Since 11,000 lbf/in is equal to 10 times 1100 lbf/in, the criterion is met and the static lateral force procedure is permitted.

### 63. Answer A

This structure has a concrete bearing wall system ($R = 4.5$) in only one direction and a special moment-resisting steel frame ($R = 8.5$) in the orthogonal direction. For design in the orthogonal direction in seismic zones 3 and 4, based on the UBC [Sec. 1630.4.3], the value of $R$ should not be greater than that used for the bearing wall system.

### 64. Answer A

Nonbuilding structures carry gravity loads and resist the effects of earthquakes. They include all self-supporting structures other than buildings.

Based on the UBC [Sec. 1634.1.6], those structures that support flexible nonstructural elements whose combined weight exceeds 25% of the weight of the structure should be designed considering the interaction effects between the structure and the supported elements.

### 65. Answer D

In designing retaining walls against overturning, per the UBC [Sec. 1611.6], the minimum resisting force should be 1.5 times the overturning moment using allowable stress design loads.

### 66. Answer D

For retaining walls, per the UBC [Sec. 1611.6], the minimum resisting force against sliding should be 1.5 times the lateral force using allowable stress design loads.

### 67. Answer B

In designing interior walls, permanent partitions, and temporary partitions where the height exceeds 6 ft (1.8 m), the UBC [Sec. 1611.5] requires the following.

- They must resist all loads to which they are subjected.

- They must resist a minimum force of 5 lbf/ft$^2$ (0.24 kN/m$^2$) applied perpendicular to the walls. The 5 psf (0.24 kN/m$^2$) load need not be applied simultaneously with wind or seismic loads.

- They must deflect a maximum of 1/240 or 1/120 of the span of walls with brittle and flexible finishes, respectively.

**68. Answer D**

Based on the UBC [Sec. 1633.2.11], all structures should be separated from adjoining structures. Separations should allow for displacement $\Delta_M$ (the maximum inelastic response displacement) due to seismic forces.

The minimum separation between buildings on the same property should be $\Delta_{MT}$ where

$$\Delta_{MT} = \sqrt{(\Delta_{MI})^2 + (\Delta_{MII})^2}$$

$\Delta_{MI}$ equals the maximum inelastic response displacement of building I, and $\Delta_{MII}$ equals the maximum inelastic response displacement of building II.

Per the UBC [Sec. 1630.9.2], $\Delta_M$ should be computed from the UBC Formula 30-17, $\Delta_M = 0.7R\Delta_S$. $\Delta_S$ is the total drift that occurs when the structure is subjected to the design forces. To determine $\Delta_M$, these drifts should be amplified.

For building I,

From UBC Table 16-N, $R$ is 4.5.

$$\begin{aligned}\Delta_{MI} &= 0.7R_I\Delta_{SI} \\ &= (0.7)(4.5)\Delta_{SI} \\ &= 3.15\Delta_{SI}\end{aligned}$$

For building II,

From UBC Table 16-N, $R$ is 5.5.

$$\begin{aligned}\Delta_{MII} &= 0.7R_{II}\Delta_{SII} \\ &= (0.7)(5.5)\Delta_{SII} \\ &= 3.85\Delta_{SII}\end{aligned}$$

Since $\Delta_{SI} = \Delta_{SII} = \Delta_S$,

$\Delta_{MI} = 3.15\Delta_S$ for building I and $\Delta_{MII} = 3.85\Delta_S$ for building II.

Use UBC Formula 33-2 to obtain the minimum required separation between the structures.

$$\begin{aligned}\Delta_{MT} &= \sqrt{(\Delta_{MI})^2 + (\Delta_{MII})^2} \\ &= \sqrt{(3.15\Delta_S)^2 + (3.85\Delta_S)^2} \\ &= \sqrt{(24.75\Delta_S)^2} \\ &= 5\Delta_S\end{aligned}$$

**69. Answer B**

*SI solution*

From UBC Table 16-N, the $R$ value for the existing structure is 8.5, and the $R$ value for the proposed structure is 4.5. Based on the UBC [Sec. 1633.2.11], adjacent buildings on the same property should be separated by at least $\Delta_{MT}$. $\Delta_{MT}$ can be obtained from UBC Formula 33-2,

$$\Delta_{MT} = \sqrt{(\Delta_{MI})^2 + (\Delta_{MII})^2}$$

$\Delta_{MI}$ and $\Delta_{MII}$ are the displacements of the adjacent buildings. For each building, solve for the maximum inelastic response displacement ($\Delta_M$). The UBC [Secs. 1630.9 and 1630.10] should be cross-referenced for this problem.

For the existing structure,

$$T = 0.8 \text{ s}$$

Based on the UBC [Sec. 1630.10.2], calculated story drift using $\Delta_M$ should not exceed 0.020 times the story height for structures having a fundamental period of 0.7 or greater.

$$0.8 \text{ s} > 0.7 \text{ s}$$

The inelastic displacement is

$$\begin{aligned}\Delta_{MI} &= (0.020)\left(\frac{18.3 \text{ m}}{5}\right)\left(\frac{100 \text{ cm}}{1 \text{ m}}\right) \\ &= 7.32 \text{ cm}\end{aligned}$$

For the proposed structure,

$$T = 0.5 \text{ s}$$

Based on the UBC [Sec. 1630.10.2], calculated story drift using $\Delta_M$ should not exceed 0.025 times the story height for structures having a fundamental period of less than 0.7 s.

$$0.5 \text{ s} < 0.7 \text{ s}$$

The inelastic displacement is

$$\begin{aligned}\Delta_{MII} &= (0.025)\left(\frac{24.4 \text{ m}}{5}\right)\left(\frac{100 \text{ cm}}{1 \text{ m}}\right) \\ &= 12.20 \text{ cm}\end{aligned}$$

For the minimum separation needed between these two buildings at a height of 18.3 m (the point of impact), use UBC Formula 33-2.

$$\Delta_{MT} = \sqrt{\begin{array}{l}\left[(7.32 \text{ cm})\left(\dfrac{18.3 \text{ m}}{3.66 \text{ m}}\right)\right]^2 \\ + \left[(12.2 \text{ cm})\left(\dfrac{24.4 \text{ m}}{4.88 \text{ m}}\right)\right]^2\end{array}}$$

$$\approx 60 \text{ cm}$$

*Customary U.S. solution*

From UBC Table 16-N, the $R$ value for the existing structure is 8.5, and the $R$ value for the proposed structure is 4.5. Based on the UBC [Sec. 1633.2.11], adjacent buildings on the same property should be separated by at least $\Delta_{MT}$. $\Delta_{MT}$ can be obtained from UBC Formula 33-2.

$$\Delta_{MT} = \sqrt{(\Delta_{MI})^2 + (\Delta_{MII})^2}$$

$\Delta_{MI}$ and $\Delta_{MII}$ are the displacements of the adjacent buildings. For each building, solve for the maximum inelastic response displacement ($\Delta_M$). The UBC [Secs. 1630.9 and 1630.10] should be cross-referenced for this problem.

For the existing structure,

$$T = 0.8 \text{ sec}$$

Based on the UBC [Sec. 1630.10.2], calculated story drift using $\Delta_M$ should not exceed 0.020 times the story height for structures having a fundamental period of 0.7 sec or greater.

$$0.8 \text{ sec} > 0.7 \text{ sec}$$

The inelastic displacement is

$$\Delta_{MI} = (0.020)\left(\frac{60 \text{ ft}}{5}\right)\left(12 \frac{\text{in}}{\text{ft}}\right)$$

$$= 2.88 \text{ in}$$

For the proposed structure,

$$T = 0.5 \text{ sec}$$

Based on the UBC [Sec. 1630.10.2], calculated story drift using $\Delta_M$ should not exceed 0.025 times the story height for structures having a fundamental period of less than 0.7 sec.

$$0.5 \text{ sec} < 0.7 \text{ sec}$$

The inelastic displacement is

$$\Delta_{MII} = (0.025)\left(\frac{80 \text{ ft}}{5}\right)\left(12 \frac{\text{in}}{\text{ft}}\right)$$

$$= 4.80 \text{ in}$$

For the minimum separation needed between these two buildings at a height of 60 ft (the point of impact), use UBC Formula 33-2.

$$\Delta_{MT} = \sqrt{\begin{array}{l}\left[(2.88 \text{ in})\left(\dfrac{60 \text{ ft}}{12 \text{ ft}}\right)\right]^2 \\ + \left[(4.80 \text{ in})\left(\dfrac{60 \text{ ft}}{16 \text{ ft}}\right)\right]^2\end{array}}$$

$$\approx 23.0 \text{ in}$$

70. Answer C

*SI solution*

From UBC Table 16-I, $Z$ is 0.3. From Table 16-K, $I_p$ is 1.0. From the UBC [Sec. 1629.3], when the soil properties are not known in sufficient detail to determine the soil profile type, type $S_D$ should be used. From UBC Table 16-Q, the seismic coefficient $C_a$ is 0.36 for seismic zone 3 and soil profile type $S_D$. From Table 16-O, the in-structure component amplification factor ($a_p$) is 1.0 and the component response modification factor ($R_p$) is 3.0. Use UBC Formula 32-2.

$$F_p = \left(\frac{a_p C_a I_p}{R_p}\right)\left(1 + (3)\left(\frac{h_x}{h_r}\right)\right)W_p$$

$$= \left(\frac{(1.0)(0.36)(1.0)}{3.0}\right)\left(1 + (3)\left(\frac{26.2 \text{ m}}{24.4 \text{ m}}\right)\right)$$

$$\times (13\,562 \text{ kg})\left(9.81 \frac{\text{m}}{\text{s}^2}\right)$$

$$= 68\,179 \text{ N}$$

Based on the UBC [Sec. 1632.2], $F_p$ should not be less than $0.7 C_a I_p W_p$.

$$F_p = (0.7)(0.36)(1.0)(13\,562 \text{ kg})\left(9.81 \frac{\text{m}}{\text{s}^2}\right)$$

$$= 33\,527 \text{ N}$$

Since $68\,179 \text{ N} > 33\,527 \text{ N}$, the criterion is met.

*Customary U.S. solution*

From UBC Table 16-I, $Z$ is 0.3. From Table 16-K, $I_p$ is 1.0. From the UBC [Sec. 1629.3], when the soil properties are not known in sufficient detail to determine the soil profile type, type $S_D$ should be used. From UBC Table 16-Q, the seismic coefficient $C_a$ is 0.36 for seismic zone 3 and soil profile type $S_D$. From Table 16-O, the in-structure component amplification factor ($a_p$) is 1.0 and the component response modification factor ($R_p$) is 3.0. Use UBC Formula 32-2.

$$F_p = \left(\frac{a_p C_a I_p}{R_p}\right)\left(1 + (3)\left(\frac{h_x}{h_r}\right)\right)W_p$$
$$= \left(\frac{(1.0)(0.36)(1.0)}{3.0}\right)\left(1 + (3)\left(\frac{86 \text{ ft}}{80 \text{ ft}}\right)\right)(30 \text{ k})$$
$$= (15.21 \text{ k})\left(1000 \frac{\text{lbf}}{\text{k}}\right)$$
$$= 15{,}210 \text{ lbf}$$

Based on the UBC [Sec. 1632.2], $F_p$ should not be less than $0.7 C_a I_p W_p$.

$$F_p = (0.7)(0.36)(1.0)(3.0 \text{ k})\left(1000 \frac{\text{lbf}}{\text{k}}\right)$$
$$= 7560 \text{ lbf}$$

Since 15,210 lbf > 7560 lbf, the criterion is met.

Also, based on the same UBC section, $F_p$ should not be more than $4 C_a I_p W_p$.

$$F_p = (4.0)(0.36)(1.0)(30 \text{ k})\left(1000 \frac{\text{lbf}}{\text{k}}\right)$$
$$= 43{,}200 \text{ lbf}$$

Since 15,210 lbf < 43,200 lbf, this criterion is also met.

## 71. Answer D

*SI solution*

From UBC Table 16-I, $Z$ is 0.4. From Table 16-K, for anchorage of machinery and equipment required for life-safety systems, the value of $I_p$ should be taken as 1.5. From the UBC [Sec. 1636.2], when the soil properties are not known in sufficient detail to determine the soil profile type, type $S_D$ should be used. From UBC Table 16-Q, the seismic coefficient $C_a$, for soil profile type $S_D$ and $Z = 0.4$ is $0.44 N_a$. From UBC Table 16-S,

the near-source factor $N_a$ for 5 k minimum distance between the site and the seismic source type A is 1.2. Therefore,

$$C_a = 0.44 N_a$$
$$= (0.44)(1.2)$$
$$= 0.53$$

Use UBC Formula 32-1.

$$F_p = 4.0 C_a I_p W_p$$
$$= (4.0)(0.53)(1.5)(1814 \text{ kg})\left(9.81 \frac{\text{m}}{\text{s}^2}\right)$$
$$= 56{,}589 \text{ N}$$

*Customary U.S. solution*

From UBC Table 16-I, $Z$ is 0.4. From Table 16-K, for anchorage of machinery and equipment required for life-safety systems, the value of $I_p$ should be taken as 1.5. From the UBC [Sec. 1636.2], when the soil properties are not known in sufficient detail to determine the soil profile types, type $S_D$ should be used. From UBC Table 16-Q, the seismic coefficient $C_a$, for soil profile type $S_D$ and $Z = 0.4$, is $0.44 N_a$. From UBC Table 16-S, the near-source factor $N_a$ for 3.11 mi minimum distance between the site and the seismic source type A is 1.2. Therefore,

$$C_a = 0.44 N_a$$
$$= (0.44)(1.2)$$
$$= 0.53$$

Use UBC Formula 32-1.

$$F_p = 4.0 C_a I_p W_p$$
$$= (4.0)(0.53)(1.5)(4000 \text{ lbf})$$
$$= 12{,}720 \text{ lbf}$$

## 72. Answer C

Intermediate moment-resisting frame (IMRF) is a concrete frame designed in accordance to the UBC [Sec. 1921.8]. Based on UBC Table 16-N, footnote 5, these systems are prohibited in seismic zones 3 and 4, except as permitted in the UBC [Sec. 1634.2]. This UBC section has an exception that indicates that IMRF systems may be used in seismic zones 3 and 4 for nonbuilding structures in occupancy categories 3 and 4 (i.e., special and standard occupancy structures) when the following criteria are met.

1) The structure is less than 50 ft (15.24 cm) in height.

2) The value of $R$ taken from UBC Table 16-N does not exceed 2.8.

## 73. Answer D

The relative lateral displacement due to seismic loads of two structures produces additional lateral force effects on the walkway. The connecting walkway should have a sliding connection on one end and a hinge on the other end to provide sufficient movement between structures.

## 74. Answer D

Plans, reports, and quantity calculations can be prepared by a registered civil engineer, his or her bona fide employees, or the owner of the firm under his or her direct supervision. However, the registered civil engineer should seal or stamp and sign all final professional documents.

## 75. Answer D

All final plans, specifications, reports, and documents should bear the seal or stamp of a registered civil engineer, as well as his or her signature.

## 76. Answer D

Geotechnical engineers should provide the analysis of geotechnical data for a site.

## 77. Answer C

The soils report (also called the *geotechnical report*) is prepared by engineers in the geotechnical branch of civil engineering. The person who signs the report and is responsible for the document must be a registered civil engineer.

## 78. Answer C

Regardless of ownership of the engineering firm, the registered civil engineer is responsible for all final plans, specifications, and reports.

## 79. Answer D

Architects, civil engineers, and structural engineers can design and sign plans of one-story residential buildings.

## 80. Answer B

By the requirements of the Division of the State Architect, State Department of General Services, registered structural engineers must sign, stamp, and seal engineering design plans for schools in California.

## 81. Answer B

Engineers should design the details of connections according to requirements and limitations prescribed in the UBC.

## 82. Answer B

The California Hospital Act requires that hospitals be fully functional and operational after an earthquake.

## 83. Answer B

The 1933 Long Beach earthquake caused much structural damage. Following this earthquake,

- the Riley Act was established, which set minimum standards for lateral force resistance in all structures

- the Field Act imparted school design approval to the Division of the State Architect, State Department of General Services

## 84. Answer D

Title 24 of the California Building Standards Administrative Code mandates that hospitals be operational after an earthquake. It also requires school buildings to resist the earthquake forces generated by major earthquakes without catastrophic collapse.

# CHAPTER 3
## DIAPHRAGM THEORY

### TOPICS

Accidental Eccentricity

Allowable Stress

Anchorage Force

Anchor Bolt

Anchor Bolt Spacing

Base Shear

Base Shear Distribution

Cantilever Pier

Cantilever Wall

Center of Mass

Center of Rigidity

Chord

Chord Force

Chord Size

Collector

Concrete Diaphragm

Deflection

Diaphragm

Diaphragm Action

Diaphragm Chord

Diaphragm Shear

Diaphragm Strut

Drag Force

Drag Strut

Drift

Eccentricity

Fixed Wall

Flexible Diaphragm

Inertia Load

Maximum Diaphragm Ratio

Minimum Base Shear

Nail Spacing

Overturning Moment

$P$-$\Delta$ Effect

Plywood Boundary Nailing

Plywood Diaphragm

Relative Rigidity

Resisting Element

Resisting Moment

Rigid Diaphragm

Rigidity

Shear Force

Shear Stress

Shear Wall Structure

Steel Deck Diaphragm

Story Drift Ratio

Story Shear

Tank

Tie-Down Force

Torsional Effect

Torsional Irregularity

Torsional Moment

Torsional Shear

Vertical Resisting Element

Walls in Parallel

Walls in Series

1. Which of the following are known as resisting elements?

    A. columns
    B. shear walls
    C. connections
    D. all of the above

2. Which of the following elements resist lateral force?

    I. shear walls
    II. horizontal diaphragms
    III. moment-resisting frames
    IV. braced frames

    A. I only
    B. II only
    C. II, III, and IV
    D. I, III, and IV

3. The base shear in a shear wall building is resisted by

    A. the parallel walls only.
    B. the perpendicular walls only.
    C. all walls.
    D. the horizontal diaphragms and all walls.

4. The transverse seismic loading is

    A. perpendicular to the shorter dimension of the building.
    B. parallel to the shorter dimension of the building.
    C. parallel to the longer dimension of the building.
    D. independent of the dimensions of the building.

5. The base shear force distribution to the floors above the base should

    A. decrease linearly with height above the base.
    B. increase linearly with height above the base.
    C. be uniform at each floor.
    D. be concentrated at the roof.

6. The walls perpendicular to the direction of lateral force

    I. are affected by ground acceleration force.
    II. have no resistance to bending moment.
    III. contribute inertia load to the diaphragm.

    A. I and II
    B. I and III
    C. II and III
    D. I, II, and III

7. For a one-story building (roof plus four walls), which walls contribute inertia load to the diaphragm shear?

    A. the walls that are perpendicular to the applied seismic force
    B. the walls that are parallel to the applied seismic force
    C. both (A) and (B)
    D. none of the walls

8. For a one-story building (roof plus four walls), which of the following elements contribute inertia load to the base shear?

    I. the roof
    II. the walls that are perpendicular to the applied seismic force
    III. the walls that are parallel to the applied seismic force

    A. I only
    B. I and II
    C. I and III
    D. I, II, and III

9. The minimum base shear calculation can be obtained from which of the following equations?

    A. $V = \left(\dfrac{2.5C_aI}{R}\right)W$
    B. $V = 0.7C_aIW$
    C. $V = 0.11C_aIW$
    D. $V = \left(\dfrac{C_vI}{RT}\right)W$

10. For the UBC equation below, what does $F_t$ represent?

$$F_x = \frac{(V - F_t)w_xh_x}{\sum_{i=1}^{n} w_ih_i}$$

    A. the stiffness of the top (roof) level
    B. the fundamental mode response
    C. the higher mode response
    D. the top story drift

11. A ten-story office building with a special moment-resisting steel frame in seismic zone 4 with near-source factor $N_v = 1.0$ is under study. The total weight of each story is 150,000 lbm (68 040 kg). Determine the base shear.

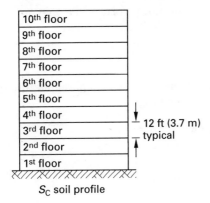

$S_C$ soil profile

A. 65,500 lbf (291 310 N)
B. 77,800 lbf (346 260 N)
C. 106,000 lbf (471 430 N)
D. 137,500 lbf (611 520 N)

12. For Problem 11, determine the concentrated force at the roof level. Assume base shear is 90,000 lbf (400 340 N).

A. 0 lbf (0 N)
B. 3200 lbf (14 230 N)
C. 8000 lbf (35 590 N)
D. 9500 lbf (42 260 N)

13. Based on the UBC, what minimum design force should the floor and roof diaphragms resist when $Z = 0.3$, $I = 1.0$, and soil profile type is $S_B$?

A. $0.15w_{px}$
B. $0.30w_{px}$
C. $0.60w_{px}$
D. $1.0w_{px}$

For Problems 14 and 15, refer to the following figure, which shows three special moment-resisting steel structures in seismic zone 4. The closest distance of these structures and known faults are 9.32 mi (15 km). These faults are not capable of producing earthquakes larger than 6.3 magnitude, and they have a low rate of seismic activity (SR = 2 mm/yr).

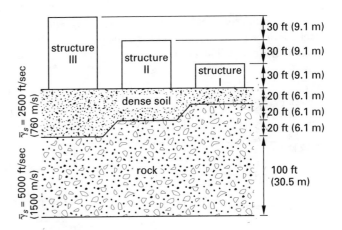

14. Which building has the largest natural period?

A. I
B. II
C. III
D. They are equal.

15. Which building has the highest seismic coefficient $(C_v)$?

A. I
B. II
C. III
D. They are equal.

16. A three-story hospital in San Francisco with a special moment-resisting steel frame has a calculated design base shear of 100,000 lbf (444 820 N). The tributary dead loads of each story including walls are given in the following illustration. What is the distributed base shear at the building roof level?

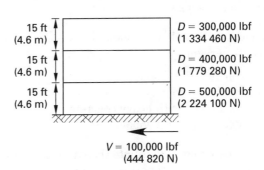

A. 33,000 lbf (146 500 N)
B. 41,000 lbf (182 000 N)
C. 49,000 lbf (217 600 N)
D. 53,000 lbf (235 200 N)

17. For a concrete shear wall building, story shears of each floor are as shown. The structure is located in seismic zone 3. What is the overturning moment at the base?

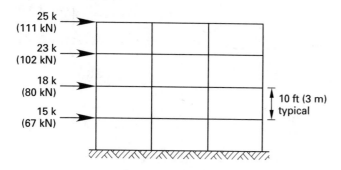

A. 810 ft-k    (1100 kN·m)
B. 1620 ft-k   (2200 kN·m)
C. 2200 ft-k   (2980 kN·m)
D. 3240 ft-k   (4390 kN·m)

18. Flexible diaphragms distribute the lateral force to the resisting elements in proportion to

A.  the relative rigidities of those elements.
B.  the tributary area of those elements.
C.  the base shear.
D.  the deflection of those elements.

19. Which of the following diaphragms are incapable of distributing torsional moments to the vertical resisting elements?

A.  flexible diaphragms
B.  rigid diaphragms
C.  both (A) and (B)
D.  diaphragms with no eccentricity

20. Which of the following diaphragms are considered flexible diaphragms?

A.  concrete slab floors
B.  reinforced concrete slab floors
C.  steel decks with heavy concrete
D.  steel decks with zonolite

21. Rigid diaphragms distribute the lateral force to the resisting elements in proportion to the

A.  relative rigidities of those elements.
B.  tributary area of those elements.
C.  base shear.
D.  deflection of those elements.

22. Which of the following statements are true regarding rigid roof diaphragms?

A.  They distribute the lateral story shears to the resisting elements.
B.  They transmit torsion to the resisting elements.
C.  They do not deform when they are subject to the lateral loads.
D.  All of the above are true.

23. Which of the following diaphragms would be considered rigid diaphragms?

A.  plywood floors
B.  wood sheathed floors
C.  light metal decks
D.  metal decks with concrete fill

24. What measure can be taken to minimize the deflection of a wood structural panel roof diaphragm?

A.  decrease the plywood thickness
B.  use double studs at the corners
C.  decrease the nail spacing
D.  increase the diaphragm span

25. Torsional effects can be critical for elements located in which of the following areas?

A.  on the center of rigidity
B.  close to the center of rigidity
C.  far from the center of rigidity
D.  none of the above

26. Which of the following elements resist torsional shear?

A.  parallel walls
B.  perpendicular walls
C.  all walls
D.  all walls and columns

27. Torsional design moment should be the moment resulting from

A.  eccentricity.
B.  accidental torsion.
C.  eccentricity and accidental torsion.
D.  bending moment.

28. A plan view of a flexible roof diaphragm is shown. The transverse loading to the diaphragm is $w$ (in units of lbf/ft (N/m)). What will the shear stresses be at lines 1 and 2 (in units of lbf/ft (N/m))? ($\rho = 1.0$)

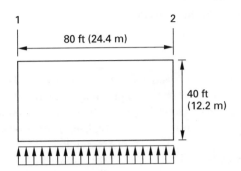

A. $0.25w$
B. $0.50w$
C. $w$
D. $2w$

29. A plan view of a one-story wood structure with a wood structural panel roof diaphragm is shown. The distributed roof diaphragm shear stress in the north-south direction is 250 lbf/ft (3648 N/m). Determine the lateral force. ($\rho = 1.0$)

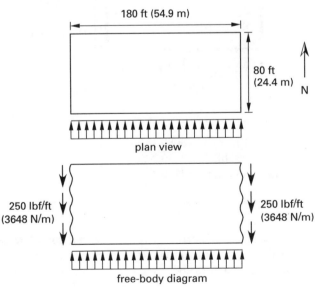

A. 110 lbf/ft   (1620 N/m)
B. 220 lbf/ft   (3240 N/m)
C. 330 lbf/ft   (4860 N/m)
D. 440 lbf/ft   (6480 N/m)

30. For the one-story wood structure shown with 400 lbf/ft (5838 N/m) diaphragm shear capacity, what should the minimum lengths of the east and west walls be? ($\rho = 1.0$)

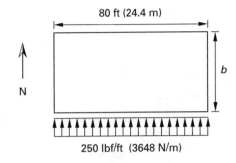

roof diaphragm
plan view

A. 12.5 ft   (3.8 m)
B. 25.0 ft   (7.6 m)
C. 37.5 ft   (11.4 m)
D. 50.0 ft   (15.2 m)

31. A one-story, wood-frame warehouse with a flexible roof diaphragm is shown in plan view. What is the length of the east-west walls if equal unit shear stress is obtained in all walls? ($\rho = 1.0$)

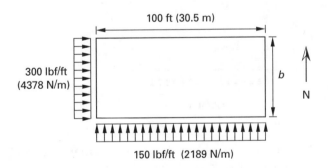

A. 25 ft   (7.6 m)
B. 35 ft   (10.7 m)
C. 50 ft   (15.2 m)
D. 70 ft   (21.3 m)

32. For loading in the north-south direction, the roof diaphragm shear capacity is 500 lbf/ft (7297 N/m). Determine $w$, the maximum distributed force on the diaphragm. ($\rho = 1.0$)

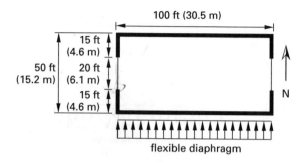

flexible diaphragm

A. 250 lbf/ft   (3640 N/m)
B. 300 lbf/ft   (4360 N/m)
C. 500 lbf/ft   (7270 N/m)
D. 750 lbf/ft   (10 910 N/m)

For Problems 33 and 34, refer to the following set of assumptions. The plan view of a one-story residential building with a wood structural panel roof diaphragm is shown. The roof has a 4 ft (1.2 m) overhang. The seismic load is in the north-south direction. ($\rho = 1.0$)

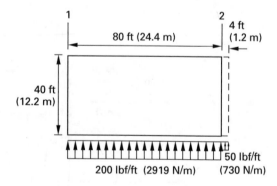

33. For the north-south direction, determine the roof diaphragm shear force at line 1.

A. 4000 lbf   (17 800 N)
B. 4200 lbf   (18 700 N)
C. 8000 lbf   (35 600 N)
D. 8200 lbf   (36 500 N)

34. For the north-south direction, determine the roof diaphragm shear force at line 2.

A. 4000 lbf   (17 790 N)
B. 4200 lbf   (18 680 N)
C. 8000 lbf   (35 590 N)
D. 8200 lbf   (36 480 N)

35. A one-story residential building with a height of 14 ft (4.3 m) is shown in plan view. This building is located in seismic zone 4. The wood structural panel roof diaphragm is well anchored to the shear walls. The plywood shear walls carry a dead load of 16 lbf/ft$^2$ (766 N/m$^2$). The uniform load shown is the seismic force on the diaphragm due to the weight of the roof and the perpendicular walls. For seismic coefficient $C_v I/RT$ use 0.1375. For the north-south direction, determine the shear wall design shear in the 20 ft (6.1 m) shear wall panel along line 2. ($\rho = 1.0$)

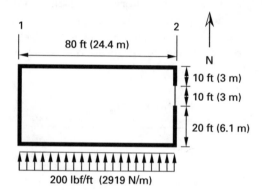

200 lbf/ft (2919 N/m)

A. 200 lbf/ft   (2930 N/m)
B. 210 lbf/ft   (3080 N/m)
C. 265 lbf/ft   (3880 N/m)
D. 280 lbf/ft   (4100 N/m)

36. Based on UBC requirements for horizontal wood structural panel diaphragms, what should be the maximum dimension ratio?

A. $2\frac{1}{2}$:1
B. 3:1
C. $3\frac{1}{2}$:1
D. 4:1

37. Based on UBC requirements for vertical wood structural panel diaphragms nailed along their edges, what should be the maximum dimension ratio in seismic zone 4?

    A. 2:1
    B. 3:1
    C. $3\frac{1}{2}$:1
    D. 4:1

38. The plan view of a structure is shown. This building has four equal sections. The structure has a wood structural panel roof diaphragm [$D = 20$ lbf/ft$^2$ (958 N/m$^2$)] and wood structural panel shear walls [$D = 25$ lbf/ft$^2$ (1197 N/m$^2$)]. The walls have a height of 16 ft (4.9 m). This structure conforms to the UBC [Sec. 1629.8.2] requirements of the simplified static lateral-force procedure. The near-source factor $N_a$ is 1.0. For the north-south direction in seismic zone 4 and soil profile type $S_C$, what should the roof diaphragm shear force be along line 1? ($\rho = 1.0$)

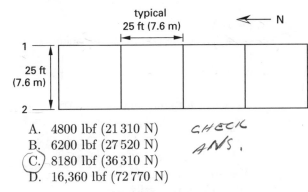

typical
25 ft (7.6 m)

←— N

1

25 ft
(7.6 m)

2

    A. 4800 lbf (21 310 N)
    B. 6200 lbf (27 520 N)
    C. 8180 lbf (36 310 N)
    D. 16,360 lbf (72 770 N)

*CHECK ANS.*

39. The wood shear wall shown is part of a one-story building in seismic zone 4 with a plywood-sheathed roof. The shear wall panels are wood structural and nailed at their edges. In the illustration, which panels comply with the UBC Maximum Diaphragm Dimension Ratios requirements?

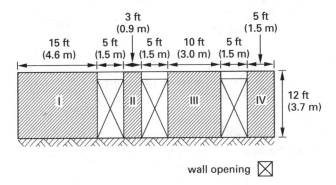

15 ft
(4.6 m)

3 ft
(0.9 m)

5 ft
(1.5 m)

5 ft
(1.5 m)

5 ft
(1.5 m)

10 ft
(3.0 m)

5 ft
(1.5 m)

I    II    III    IV

12 ft
(3.7 m)

wall opening ⊠

    A. I and III
    B. II and IV
    C. I, II, and IV
    D. I, III, and IV

For Problems 40 through 42, refer to the following set of assumptions. A wood structure with a wood structural panel roof diaphragm is shown in plan view. Two retail stores are separated by a common wood shear wall parallel to the north-south lateral loading. The height of the building is 14 ft (4.3 m). The roof diaphragm is adequately anchored to the shear walls. ($\rho = 1.0$)

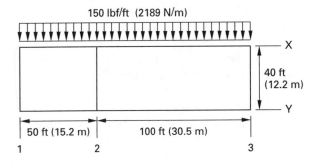

north-south loading

150 lbf/ft (2189 N/m)

X

40 ft
(12.2 m)

Y

50 ft (15.2 m)    100 ft (30.5 m)

1    2    3

40. Determine the roof diaphragm shear force along line 3.

    A. 3800 lbf (16 910 N)
    B. 5600 lbf (24 920 N)
    C. 7500 lbf (33 380 N)
    D. 9000 lbf (40 050 N)

41. Along line X between lines 1 and 2, determine the maximum diaphragm chord force in the chord member.

    A. 750 lbf (3240 N)
    B. 1200 lbf (5180 N)
    C. 2300 lbf (9930 N)
    D. 4700 lbf (20 300 N)

42. Determine the diaphragm chord force at the intersection of lines Y and 1.

    A. 0 lbf (0 N)
    B. 290 lbf (1290 N)
    C. 600 lbf (2670 N)
    D. 1200 lbf (5340 N)

*FLEXIBLE DIAPHRAGM SHEAR*

43. A wood-frame shear wall is shown in elevation. The shear wall is part of a one-story structure with a flexible roof diaphragm. The shear wall panels have the same thickness. What is the approximate lateral load carried by panel I?

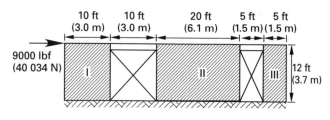

wall opening ☒

A. 1800 lbf  (7840 N)
B. 2600 lbf  (11 330 N)
C. 3000 lbf  (13 070 N)
D. 9000 lbf  (39 220 N)

44. What is the purpose of providing drag struts (collectors) at the diaphragm boundaries?

A. to resist moments
B. to transmit unsupported horizontal diaphragm shear to the supporting shear walls
C. to distribute torsion
D. none of the above

45. The force in a drag strut is

A. tension only.
B. compression only.
C. either tension or compression.
D. none of the above.

46. The wood structural panel roof diaphragm of a one-story building is shown in plan view. The north shear wall's length is not the full width of the building. Determine the location of the maximum strut force. ($\rho = 1.0$)

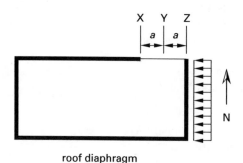

roof diaphragm

A. at point X
B. at point Y (at the middle of the opening)
C. at point Z
D. at both points X and Z

47. A wood structural panel roof diaphragm is shown in plan view. The left wall has an opening at its center. For the loading shown, where in the drag strut does the maximum force occur? ($\rho = 1.0$)

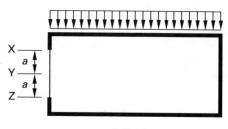

roof diaphragm

A. at point X
B. at point Y (at the middle of the opening)
C. at point Z
D. at both points X and Z

48. Choose the correct statements for the strut force.

I. The strut force in compression will remain a compression force if the direction of the lateral force is reversed.
II. The strut forces are zero at the outside ends of the walls.
III. The greater magnitude of the strut force or the chord force at corresponding points along the wall determines the critical loading condition.

A. I and II
B. I and III
C. II and III
D. I, II, and III

49. What is the purpose of providing chords at diaphragm boundaries?

A. to distribute torsion
B. to distribute lateral load
C. to resist moments
D. none of the above

50. Which of the following materials would be suitable to use as diaphragm chords?

    A. wood
    B. steel
    C. masonry
    D. all of the above

51. The force in the chord member is obtained from which of the following formulas?

    A. force in chord $= \dfrac{M}{b}$

    B. force in chord $= \dfrac{wL^2}{8b}$

    C. force in chord $= \dfrac{FL}{8b}$

    D. all of the above

52. For a flexible roof diaphragm adequately anchored to the shear walls, which of the following statements identifies the correct chord forces at lines 1 and 2? ($\rho = 1.0$)

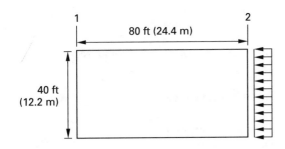

    A. The maximum chord forces at lines 1 and 2 are equal.
    B. The maximum chord force at line 2 is twice the chord force at line 1.
    C. The chord force at line 1 is generally ignored.
    D. The total chord force at lines 1 and 2 is $40w$.

53. A residential one-story wood structure with a wood structural panel roof diaphragm is shown in plan view. The lateral force is in the north-south direction. The calculated roof diaphragm shear capacity is 200 lbf/ft (2919 N/m) in the direction of loading. The roof diaphragm is adequately anchored to the shear walls. What will be the maximum force in the chord members? ($\rho = 1.0$)

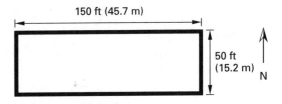

    A. 7500 lbf  (33 350 N)
    B. 11,000 lbf  (48 910 N)
    C. 12,700 lbf  (56 470 N)
    D. 14,300 lbf  (63 590 N)

54. A one-story retail store, a wood structure with a wood structural panel roof diaphragm, is shown in plan view. At the intersection of lines X and 1, determine the force in the chord member. ($\rho = 1.0$)

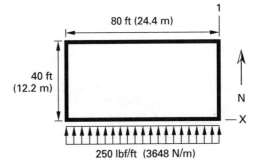

    A. 0 lbf  (0 N)
    B. 1250 lbf  (5560 N)
    C. 2500 lbf  (11 120 N)
    D. 5000 lbf  (22 240 N)

55. A one-story, wood-frame commercial building has a wood structural panel roof diaphragm, and its south wall has a 40 ft (12.2 m) opening. Determine the chord force at the intersection of lines X and 1. ($\rho = 1.0$)

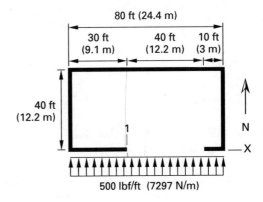

A. 2350 lbf (10 340 N)
B. 4700 lbf (20 690 N)
C. 9400 lbf (41 370 N)
D. 11,750 lbf (51 710 N)

56. Which of the following statements regarding diaphragm chord members is incorrect?

    I. The maximum chord force occurs at the location of the maximum moment.
    II. The maximum chord force occurs at the chord ends.
    III. The minimum chord force is zero.

A. II
B. III
C. I and III
D. I, II, and III

57. The plan view of a wood-frame one-story retail store is shown. The south wall has a 35 ft (10.7 m) shear wall panel. How long should this shear wall panel be to have an equal magnitude drag and chord force? ($\rho = 1.0$)

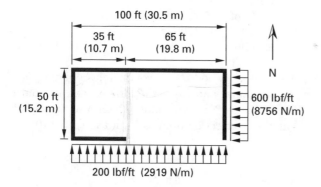

A. 0 ft (0 m)
B. 40 ft (12.2 m)
C. 60 ft (18.3 m)
D. 75 ft (22.9 m)

58. The wood structural panel roof diaphragm of a one-story building is shown in plan view. The walls are of wood-frame construction. For the south wall, at what point are the magnitudes of the drag strut and the chord force identical? ($\rho = 1.0$)

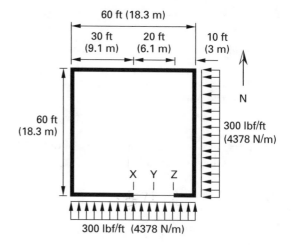

A. at point X
B. at point Y (at the middle of the opening)
C. at point Z
D. none

59. Drift depends on

A. the story height.
B. the building height.
C. the shear load.
D. all of the above.

60. The $P$-$\Delta$ effect on shears, axial forces, and moments of frame members is considered as

A. the primary moment.
B. the secondary moment.
C. the overall moment.
D. none of the above.

61. For the one-story building shown, determine the story drift ratio.

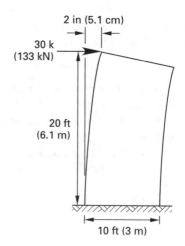

A. 0.008
B. 0.1
C. 3.0
D. 120.0

62. When the overturning moment increases faster than the restoring moment from the frame stiffness, the frame is considered to be

A. stable.
B. unstable.
C. X-braced.
D. a thick shear wall.

63. Using the appropriate UBC table, what should be the maximum allowable shear stress for a wood structural panel roof diaphragm when the following data apply?

- $\frac{3}{8}$ in (10 mm) thick Structural I plywood
- 2 in (51 mm) framing member
- Case 1—Blocked
- 8d common nails
- nail spacing at diaphragm boundaries is 4 in (102 mm) O.C.
- nail spacing at other panel edges is 6 in (152 mm) O.C.
- field nail spacing is 12 in (30 cm) O.C.

A. 320 lbf/ft   (4670 N/m)
B. 360 lbf/ft   (5250 N/m)
C. 400 lbf/ft   (5840 N/m)
D. 530 lbf/ft   (7730 N/m)

64. In accordance with the appropriate UBC table, for a single-family dwelling, determine the nail spacing at the diaphragm boundaries of a wood structural panel roof diaphragm when the following data apply.

- 15/32 in (12 mm) thick Structural I panels
- 3 in (76 mm) framing member
- Case 2—Blocked
- 10d common nails

- 690 lbf/ft (10 070 N/m) shear force (calculated)
- 12 in (30 cm) O.C. field nailing

A. 2.0 in   (51 mm)
B. 2.5 in   (64 mm)
C. 4.0 in   (102 mm)
D. 6.0 in   (152 mm)

65. An engineering firm calculated that the maximum shear force in the wood structural panel shear walls of a one-story residential building will be 380 lbf/ft (5546 N/m). The building has a flexible roof diaphragm. What should be the nail spacing at the wood structural panel edges of the shear walls? The following data apply.

- 15/32 in (12 mm) thick Structural I panels
- 8d common nails
- wood structural panels applied directly to the framing members
- blocking required
- 12 in (30 cm) O.C. field nailing

A. 2 in   (51 mm)
B. 3 in   (76 mm)
C. 4 in   (102 mm)
D. 6 in   (152 mm)

66. The resisting moment of the parallel walls of a one-story building is less than the overturning moment in the direction of the applied lateral force. What should the proper action be?

A. anchor the wall panels to the foundation
B. increase the weight of the walls
C. increase the length of the walls
D. all of the above

67. The wood structural panel shear wall shown in elevation is the south wall of a one-story building with a plywood roof diaphragm. Determine the required number of $\frac{3}{4}$ in (19 mm) diameter anchor bolts necessary to transfer the 11,000 lbf (48 930 N) seismic load to the foundation from the wall plate. These bolts should be in addition to the anchor bolts used for the uplift

anchorage. Ignore the roof and wall dead loads. Use UBC Table 21-N for allowable shear on bolts for grouted masonry.

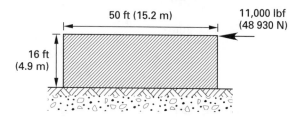

A. 2

B. 4

C. 6

D. 10

68. The east wall of a two-story, wood-frame building is shown in elevation. The lateral loads from the horizontal roof and second-floor diaphragms are 8000 lbf (35 586 N) and 5000 lbf (22 241 N), respectively. The dead load for the shear walls is 16 lbf/ft² (766 N/m²). Ignore the roof dead load contribution to the shear walls. What should the anchorage force be at location X between the two stories? The building is located in seismic zone 4.

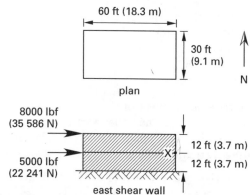

A. 320 lbf   (1570 N)

B. 610 lbf   (2870 N)

C. 1000 lbf   (4680 N)

D. Anchorage is not needed.

69. The one-story storage building in seismic zone 4 is 14 ft (4.3 m) in height. The concrete shear walls are 8 in (20.3 cm) thick. The wood structural panel roof is 15/32 in (12 mm) STR I. The blocked nails are 10d common, and the ledgers are a nominal 3 in (76 mm)

thick (Douglas fir-larch). The diaphragm, ledger, and walls are anchored. The roof shear capacity per unit length along the north-south walls is 600 lbf/ft (8756 N/m). Assume the shear loading is parallel to the grain. Based on UBC requirements, determine the required size of ledger bolts necessary to resist the roof shear force if the bolts are spaced every 4 ft (1.2 m).

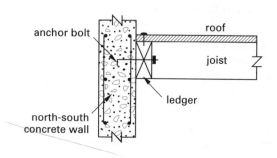

A. $\frac{5}{8}$ in   (16 mm)

B. $\frac{3}{4}$ in   (19 mm)

C. $\frac{7}{8}$ in   (22 mm)

D. 1 in   (25 mm)

70. What is the relationship between the rigidity, $R$, and the deflection, $\Delta$, produced by a unit load?

A. $\Delta = R$

B. $\Delta = \dfrac{1}{R}$

C. $\Delta = R^2$

D. $\Delta = \dfrac{1}{R^2}$

71. The rigidity of a wall depends on the

A. the modulus of elasticity.

B. the height.

C. the thickness.

D. all of the above.

72. For fixed and cantilever piers, how are the rigidities and deflections affected by increasing the ratio of height to width?

A. The rigidities will increase while the deflections decrease.

B. The rigidities will decrease while the deflections increase.

C. Both will increase.

D. Both will decrease.

73. A one-story building with a rigid diaphragm is supported by masonry shear wall panels. The north wall is shown. The wall has a uniform thickness and modulus of elasticity. It is necessary to divide the wall into piers and beams to determine the rigidity of this wall with openings. How many piers does this wall have?

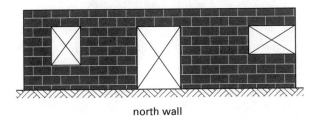

north wall

wall opening ⊠

A. two
B. three
C. four
D. six

74. The south wall of a masonry shear wall building with a rigid roof diaphragm is shown. The wall consists of panels I and II. The walls have a uniform thickness and modulus of elasticity. What will be the total deflection of this wall due to a unit load?

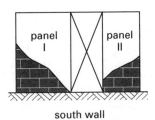

south wall

wall opening ⊠

A. $\Delta_{\text{total}} = \Delta_{\text{I}} + \Delta_{\text{II}}$

B. $\Delta_{\text{total}} = \dfrac{1}{\Delta_{\text{I}}} + \dfrac{1}{\Delta_{\text{II}}}$

C. $\Delta_{\text{total}} = \dfrac{\Delta_{\text{I}}\Delta_{\text{II}}}{\Delta_{\text{I}} + \Delta_{\text{II}}}$ *PARALLEL*

D. $\Delta_{\text{total}} = \dfrac{\Delta_{\text{I}} + \Delta_{\text{II}}}{\Delta_{\text{I}}\Delta_{\text{II}}}$

75. The masonry wall panels shown have the same thickness and modulus of elasticity. Assuming panels I and II are fixed at the top, what will the total deflection be? Ignore rotational effects.

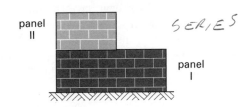

*SERIES*

panel II

panel I

A. $\Delta_{\text{total}} = \Delta_{\text{I}}\Delta_{\text{II}}$

B. $\Delta_{\text{total}} = \Delta_{\text{I}} + \Delta_{\text{II}}$

C. $\dfrac{1}{\Delta_{\text{total}}} = \Delta_{\text{I}} + \Delta_{\text{II}}$

D. $\dfrac{1}{\Delta_{\text{total}}} = \dfrac{1}{\Delta_{\text{I}}} + \dfrac{1}{\Delta_{\text{II}}}$

76. The north and south walls of a warehouse are shown in elevation. All walls have the same thickness and modulus of elasticity. Assume that the walls are fixed at the top and that the warehouse has a rigid roof diaphragm. Which of the following statements are correct?

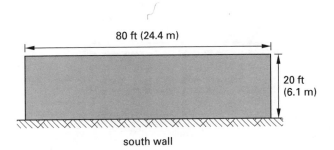

80 ft (24.4 m)

20 ft (6.1 m)

south wall

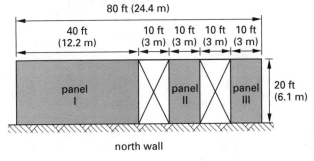

80 ft (24.4 m)

40 ft (12.2 m) | 10 ft (3 m) | 10 ft (3 m) | 10 ft (3 m) | 10 ft (3 m)

panel I | panel II | panel III

20 ft (6.1 m)

north wall

wall opening ⊠

A. The rigidity of the south wall is greater.
B. The deflection of the north wall is greater.
C. Generally, openings reduce the rigidity of walls.
D. All of the above are true.

77. The masonry walls shown have the same thickness and modulus of elasticity. Assume the walls are fixed at the top. Use $t = 1$ in (25 mm), $F = 100,000$ lbf (444 820 N), and $E = 1 \times 10^6$ lbf/in$^2$ (6.9 × 10$^6$ kPa). Which wall has the highest deflection?

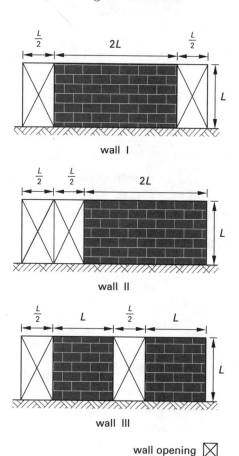

wall I

wall II

wall III

wall opening ⊠

A. wall I
B. wall II
C. wall III
D. The deflections are equal.

78. For the masonry wall shown, the thickness and modulus of elasticity are uniform. Assume the wall is fixed at the top and bottom. Use $t = 1$ in (25 mm), $F = 100,000$ lbf (444 820 N), and $E = 1 \times 10^6$ lbf/in$^2$ (6.9 × 10$^6$ kPa). Determine the relative rigidity.

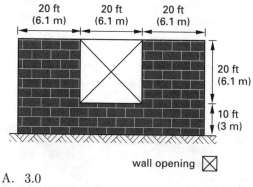

wall opening ⊠

A. 3.0
B. 4.0
C. 5.0
D. 6.0

79. The masonry wall shown has a uniform thickness and modulus of elasticity. Assume the wall is fixed at the top. Use $t = 1$ in (25 mm), $F = 100,000$ lbf (444 820 N), and $E = 1 \times 10^6$ lbf/in$^2$ (6.9 × 10$^6$ kPa). What is the rigidity of the shear wall?

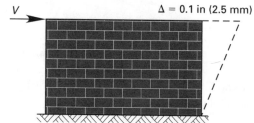

A. 0.1
B. 1.0
C. 3.0
D. 10.0

80. The wall shown is fixed at the top. The rigidity of the entire wall is 10.0, and the deflection for the steel frame is 1 in (25 mm). Determine the rigidities of the steel frame and the masonry shear wall.

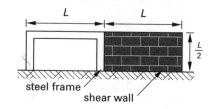

A. 5.0 and 5.0, respectively
B. 9.1 and 0.9, respectively
C. 1.0 and 9.0, respectively
D. 9.0 and 1.0, respectively

81. The south wall of a one-story building with a rigid roof diaphragm is shown. The rigidity of the steel frame is nine times the rigidity of the shear wall. Determine the lateral loads carried by the steel frame and the shear wall.

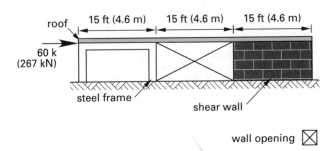

wall opening $\boxtimes$

   A.  54 k and 6 k (240 kN and 27 kN), respectively
   B.  46 k and 14 k (205 kN and 62 kN), respectively
   C.  40 k and 20 k (180 kN and 90 kN), respectively
   D.  30 k and 30 k (130 kN and 130 kN), respectively

82. The masonry south wall shown is a part of a one-story warehouse with a rigid roof diaphragm. The south wall has two panels with equal rigidities. Which panel carries the larger load?

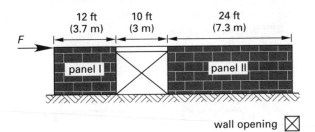

wall opening $\boxtimes$

   A.  Panel I carries the larger load.
   B.  Panel II carries the larger load.
   C.  Panel II carries twice the load of panel I.
   D.  Both panels carry an equal load.

83. The west wall of a one-story building with a rigid roof diaphragm is shown. The masonry shear panels have the same thickness and modulus of elasticity. Assume that the properties for the steel frames are the same. Determine the lateral load carried by the masonry shear panels.

$t = 1$ in (25 mm)
$F = 100{,}000$ lbf (444 820 N)
$E = 1 \times 10^6$ lbf/in$^2$ (6.9 × 10$^6$ kPa)

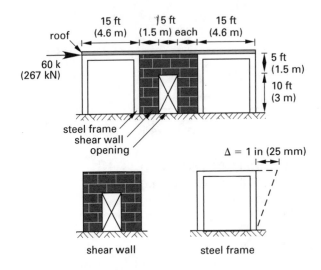

   A.  13 k  (58 kN)
   B.  19 k  (85 kN)
   C.  22 k  (98 kN)
   D.  30 k  (133 kN)

84. The base shear is distributed to the four stories of a building. The story heights are equal. The total height of the building is 48 ft (14.6 m). There is no additional force applied to the roof level. The distributed forces increase linearly with the height above the base as follows.

$$1^{\text{st}} \text{ floor} = 20{,}000 \text{ lbf} \quad (88\,960 \text{ N})$$
$$2^{\text{nd}} \text{ floor} = 45{,}000 \text{ lbf} \quad (200\,170 \text{ N})$$
$$3^{\text{rd}} \text{ floor} = 71{,}000 \text{ lbf} \quad (315\,820 \text{ N})$$
$$4^{\text{th}} \text{ floor} = 93{,}000 \text{ lbf} \quad (413\,680 \text{ N})$$

Determine the overturning moment at the base.

   A.  2750 ft-k  (3760 kN·m)
   B.  4600 ft-k  (6300 kN·m)
   C.  6550 ft-k  (8970 kN·m)
   D.  8350 ft-k  (11 440 kN·m)

85. The wood structural panel shear wall shown in elevation is part of a wood structure with a wood structural panel roof diaphragm. The roof diaphragm is adequately anchored to the shear walls. The seismic

coefficient, $C_v I / RT$, is 0.1375, and the wall dead load is 16 lbf/ft² (766 N/m²). What shear capacity is needed in the shear wall?

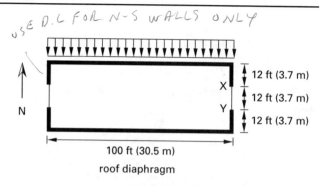

USE D.L FOR N-S WALLS ONLY

12 ft (3.7 m)
12 ft (3.7 m)
12 ft (3.7 m)
100 ft (30.5 m)

roof diaphragm

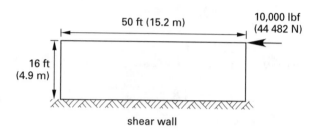

50 ft (15.2 m)

10,000 lbf (44 482 N)

16 ft (4.9 m)

shear wall

A. 200 lbf/ft (2920 N/m)
B. 218 lbf/ft (3180 N/m)
C. 235 lbf/ft (3430 N/m)
D. 270 lbf/ft (3940 N/m)

86. The wood structural panel shear wall shown in elevation is part of a one-story wood structure. The wall dead load is 20 lbf/ft² (958 N/m²). Ignore the roof dead load contribution to uplift resistance. What is the resisting moment provided by the weight of the wall against overturning?

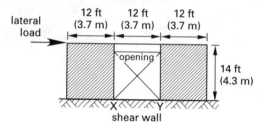

lateral load

12 ft (3.7 m)  12 ft (3.7 m)  12 ft (3.7 m)

opening

14 ft (4.3 m)

X  Y

shear wall

A. 1540 lbf (6640 N)
B. 2500 lbf (10 780 N)
C. 3950 lbf (17 030 N)
D. 5680 lbf (27 230 N)

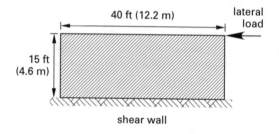

40 ft (12.2 m)

lateral load

15 ft (4.6 m)

shear wall

A. 102,000 ft-lbf (139 380 N·m)
B. 120,000 ft-lbf (163 980 N·m)
C. 204,000 ft-lbf (278 760 N·m)
D. 216,000 ft-lbf (295 160 N·m)

87. The plan view of a one-story office building in San Francisco is as shown. The roof is flexible wood structural panel and the shear walls are wood structural panels. The roof dead load is 20 lbf/ft² (958 N/m²), and the wall dead loads are 16 lbf/ft² (766 N/m²). There is no geotechnical data for the site. Use UBC simplified design base shear procedure with near-source factor $N_a = 1.0$. What are the tie-down forces at points X and Y? ($\rho = 1.0$)

88. The plan view of a rigid roof diaphragm is shown. A lateral load of 500 lbf/ft (7297 N/m) in the north-south direction is acting on the building. Assume the building is stable. The rigidities for the end shear walls are given. Ignore torsional and eccentricity effects. Determine how much of the lateral force will be resisted by the 20 ft (6.1 m) wall. ($\rho = 1.0$)

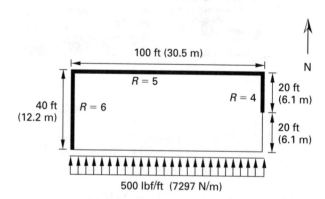

100 ft (30.5 m)

R = 5
R = 6
R = 4

40 ft (12.2 m)

20 ft (6.1 m)
20 ft (6.1 m)

N

500 lbf/ft (7297 N/m)

A. 10,000 lbf (44 500 N)
B. 20,000 lbf (89 000 N)
C. 30,000 lbf (133 500 N)
D. 40,000 lbf (178 000 N)

For Problems 89 and 90, refer to the following set of assumptions.

A car-wash building with a rigid roof diaphragm is shown in plan view. Three concrete shear walls parallel to the direction of the north-south lateral force are indicated along with their rigidities. Ignore effects of torsion and eccentricity. Assume the building is stable. ($\rho = 1.0$)

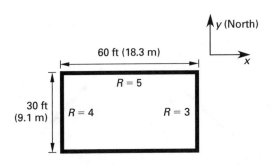

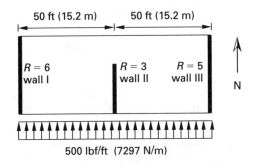

89. Determine how much of the lateral force will be resisted by wall I.

    A. 10,700 lbf  (47 480 N)
    B. 16,700 lbf  (74 100 N)
    C. 17,800 lbf  (78 980 N)
    D. 21,500 lbf  (95 400 N)

90. Which wall will resist the largest proportion of the north-south lateral force?

    A. wall I
    B. wall II
    C. wall III
    D. All walls resist equally.

91. The concrete walls of a one-story building with a rigid roof diaphragm are 12 ft (3.7 m) in height. The rigidities of the north, east, and west walls are indicated in the building plan view. Do not account for uncertainties in the locations of loads. What should be the rigidity of the south wall if the center of rigidity in the $y$-direction coincides with the center of the diaphragm?

    A. 2
    B. 4
    C. 5
    D. 7

92. The rigid roof diaphragm of a factory is shown in plan view. The relative rigidities of the concrete walls are given. Determine the location of the center of rigidity in the $x$- and $y$-directions. Use the lower left corner, point O, as the origin.

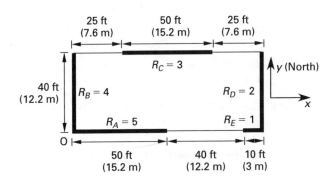

    A. 25 ft and 15 ft  (7.6 m and 4.6 m), respectively
    B. 30 ft and 20 ft  (9.1 m and 6.1 m), respectively
    C. 33 ft and 13 ft  (10.0 m and 4.0 m), respectively
    D. 50 ft and 20 ft  (15.2 m and 6.1 m), respectively

93. A one-story octagonal building is shown in plan view. The concrete roof and shear walls have a uniform thickness throughout. The walls' rigidities are given. Determine the location of the center of rigidity in the $x$- and $y$-directions. Use the lower left corner, point O, as the origin.

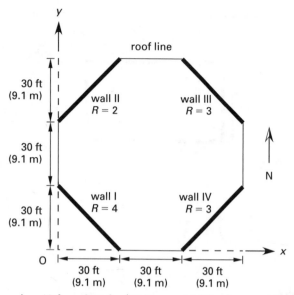

roof line

wall II
R = 2

wall III
R = 3

wall I
R = 4

wall IV
R = 3

30 ft
(9.1 m)

30 ft
(9.1 m)

30 ft
(9.1 m)

O

30 ft
(9.1 m)

30 ft
(9.1 m)

30 ft
(9.1 m)

N

A.  40 ft and 42 ft  (12.2 m and 12.8 m), respectively
B.  42 ft and 40 ft  (12.8 m and 12.2 m), respectively
C.  45 ft and 40 ft  (13.7 m and 12.2 m), respectively
D.  45 ft and 45 ft  (13.7 m and 13.7 m), respectively

94. The accidental eccentricity of a building has been calculated as 4 ft (1.2 m) for the direction of loading under consideration. What is the length of this building, perpendicular to the direction of loading?

A.  20 ft  (6.0 m)
B.  40 ft  (12.0 m)
C.  80 ft  (24.0 m)
D.  104 ft  (32.0 m)

95. The plan of a one-story building is shown. The base shear is 50,000 lbf (222 410 N) for the north-south direction. The building has a rigid roof diaphragm with concrete shear walls. The location of the centers of rigidity and mass as computed from the southwest corner are given below. What is the maximum positive torsional moment in the north-south direction?

|  | center of mass |  | center of rigidity |
|---|---|---|---|
| $\overline{x}$ | 40 ft (12.2 m) | $\overline{x}_R$ | 38 ft (11.6 m) |
| $\overline{y}$ | 12.5 ft (3.8 m) | $\overline{y}_R$ | 9.5 ft (2.9 m) |

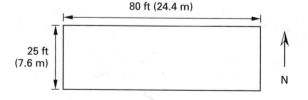

80 ft (24.4 m)

25 ft
(7.6 m)

N

A.  100,000 ft-lbf  (133 450 N·m)
B.  150,000 ft-lbf  (200 170 N·m)
C.  300,000 ft-lbf  (400 340 N·m)
D.  350,000 ft-lbf  (467 060 N·m)

96. The plan view of a one-story masonry shear wall structure with a rigid roof diaphragm is shown. The locations of the centers of mass and rigidity are illustrated. Neglect accidental eccentricity.

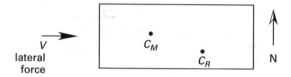

V
lateral
force

$C_M$

$C_R$

N

The direction of the torsional moment on the diaphragm for the lateral force direction shown is

A.  clockwise.
B.  counterclockwise.
C.  opposite the direction of lateral force.
D.  none of the above.

97. The plan view of a one-story masonry shear wall structure with a rigid roof diaphragm is shown. The relative rigidity of each shear wall, the center of mass, the center of rigidity, and the base shear for the north-south direction are given in the illustration. What is the lateral force in the north wall due to torsion?

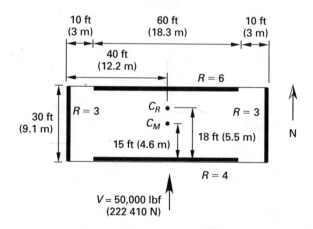

10 ft
(3 m)

60 ft
(18.3 m)

10 ft
(3 m)

40 ft
(12.2 m)

R = 6

30 ft
(9.1 m)

R = 3

$C_R$
$C_M$

R = 3

15 ft (4.6 m)

18 ft (5.5 m)

R = 4

V = 50,000 lbf
(222 410 N)

N

A.  780 lbf  (3370 N)
B.  1220 lbf  (5280 N)
C.  2000 lbf  (8650 N)
D.  2800 lbf  (12 110 N)

98. Consider a one-story masonry shear wall structure with a rigid roof diaphragm. The lateral force is in the north-south direction. The centers of mass and rigidity are as shown. The building is acted upon by a torsional moment. The rotation about the center of rigidity is resisted by a torsional shear stress in all walls. For walls I, II, III, and IV, choose the direction of the torsional shear that resists the torsional moment.

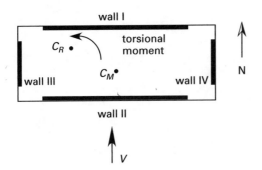

A. east, east, south, and north, respectively
B. east, west, north, and south, respectively
C. west, east, south, and north, respectively
D. west, west, north, and south, respectively

99. The plan view of a one-story masonry shear wall structure with a rigid roof diaphragm is shown. The relative rigidity of each shear wall, the center of mass, and the center of rigidity are given in the illustration. The base shear is 60,000 lbf (266 900 N) in the north-south direction. What is the total force in the west wall?

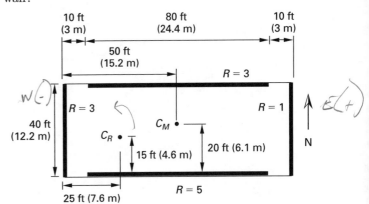

A. 22,500 lbf (100 100 N)
B. 36,400 lbf (162 000 N)
C. 45,000 lbf (200 170 N)
D. 52,000 lbf (231 310 N)

# CHAPTER 3
## Solutions

In solutions 28–35, 38, 40–42, 46, 47, 52–55, 57, 58, and 87–90, $\rho$ represents a reliability/redundancy factor that should be assigned to all structures according to the UBC [Sec. 1630.1.1]. This factor is based on the extent of structural redundancy and its lateral force-resisting systems.

**1. Answer D**
Elements that resist story shears are known as *resisting elements*. They are columns, shear walls, connections, and structural systems, such as braced frames or moment-resisting frames.

**2. Answer D**
Resisting elements (e.g., columns, shear walls, braced frames, moment-resisting frames, and connections) resist lateral forces. Horizontal diaphragms (e.g., floors and ceilings) distribute lateral forces to the resisting elements.

**3. Answer A**
The base shear is the total force delivered to the structure at the base. In perpendicular walls, the forces are tensile and compressive, while in parallel walls, the forces are shear. Therefore, in shear wall buildings, parallel walls resist the base shear.

**4. Answer B**
The direction of seismic loading applied to a structure is typically either in the transverse (i.e., short dimension) or longitudinal (i.e., long dimension) direction. Therefore, transverse seismic loading describes a loading that is parallel to the shorter dimension of the structure.

**5. Answer B**
Per the UBC [Sec. 1630.5], base shear forces should be distributed throughout the height of structures in conformance with UBC Formulas 30-13, 30-14, and 30-15. At each level, the distributed force, $F_x$, increases linearly as the height above the base increases.

**6. Answer B**
Walls that are perpendicular to the lateral force are affected by ground acceleration forces and carry all tensile and compressive forces. They resist the bending moment. Perpendicular walls transfer their own inertia forces to the base of the structure.

**7. Answer A**
The walls perpendicular to the direction of the seismic forces resist the bending moment, and they also contribute inertia load to the diaphragm shear.

**8. Answer D**
All of the inertia forces resulting from the mass of the structure along with all masses attached to or contained by the structure should be carried from their origins to the foundations. Roof diaphragms transmit inertia forces from their own weight plus the weight of the perpendicular walls to the vertical resisting elements. The vertical shear-resisting elements are the parallel walls that transfer forces from the roof diaphragms, plus their own inertia forces, to the base of structures.

**9. Answer C**
Based on the UBC [Sec. 1630.2.1], the minimum total design base shear should be obtained from UBC Formula 30-6, $V = 0.11C_a IW$.

In addition, for seismic zone 4, the minimum total design base shear should be calculated from UBC Formula 30-7, $V = (0.8ZN_v I/R)W$.

**10. Answer C**
UBC Formula 30-15 distributes the base shear force to each story in accordance with the distribution of mass at that level. $F_t$ represents a concentrated force at the top level (i.e., the roof) that is a supplemental force to $F_x$ at that level. $F_t$ accounts for the response to higher modes, which is more significant for longer period

structures. Based on the UBC [Sec. 1630.5], $F_t$ should be determined from UBC Formula 30-14, $F_t = 0.07TV$.

## 11. Answer B

*SI solution*

For seismic zone 4, from UBC Table 16-I, $Z$ is 0.4. For an office building, from UBC Table 16-K, $I$ is 1.0. For a special moment-resisting steel frame, from UBC Table 16-N, $R$ is 8.5. Per the UBC [Sec. 1630.2.2, Item 1], $C_t$ is 0.0853.

$$h_n = (10 \text{ floors})(3.7 \text{ m})$$
$$= 37 \text{ m}$$
$$T = C_t(h_n)^{3/4}$$
$$= (0.0853)(37 \text{ m})^{3/4}$$
$$= 1.27 \text{ s}$$

Based on UBC Table 16-R, the seismic response coefficient $C_v$ is $0.56N_v$ for soil profile type $S_C$ and $Z = 0.4$. Therefore,

$$C_v = (0.56)(1.0)$$
$$= 0.56$$
$$W = (10 \text{ floors})(68\,040 \text{ kg})$$
$$= 680\,400 \text{ kg}$$

Use UBC Formula 30-4.

$$V = \left(\frac{C_vI}{RT}\right)W$$
$$= \left(\frac{(0.56)(1.0)}{(8.5)(1.27)}\right)(680\,400 \text{ kg})\left(9.81 \frac{\text{m}}{\text{s}^2}\right)$$
$$= 346\,257 \text{ N}$$

*Customary U.S. solution*

For seismic zone 4, from UBC Table 16-I, $Z$ is 0.4. For an office building, from UBC Table 16-K, $I$ is 1.0. For a special moment-resisting steel frame, from UBC Table 16-N, $R$ is 8.5. Per UBC [Sec. 1630.2.2, Item 1], $C_t$ is 0.035.

$$h_n = (10 \text{ floors})(12 \text{ ft})$$
$$= 120 \text{ ft}$$
$$T = C_t(h_n)^{3/4}$$
$$= (0.035)(120 \text{ ft})^{3/4}$$
$$= 1.27 \text{ sec}$$

Based on UBC Table 16-R, the seismic response coefficient $C_v$ is $0.56N_v$ for soil profile type $S_C$ and $Z = 0.4$.

Therefore,

$$C_v = (0.56)(1.0)$$
$$= 0.56$$
$$W = (10 \text{ floors})(150,000 \text{ lbf})$$
$$= 1,500,000 \text{ lbf}$$

Use UBC Formula 30-4.

$$V = \left(\frac{C_vI}{RT}\right)W$$
$$= \left(\frac{(0.56)(1.0)}{(8.5)(1.27)}\right)(1,500,000 \text{ lbf})$$
$$= 77,814 \text{ lbf}$$

## 12. Answer C

*SI solution*

From the solution to Problem 11, $T = 1.27$ s. Since 1.27 s > 0.7 s, use the UBC [Sec. 1630.5], Formula 30-14.

$$F_t = 0.07TV$$
$$= (0.07)(1.27 \text{ s})(400\,340 \text{ N})$$
$$= 35\,590 \text{ N}$$

Check that $F_t \le 0.25V$.

$$0.25V = (0.25)(400\,340 \text{ N})$$
$$= 100\,085 \text{ N}$$

Since $35\,590$ N $< 100\,085$ N, the value of $35\,590$ N is used for $F_t$.

*Customary U.S. solution*

From the solution to Problem 11, $T = 1.27$ sec. Since 1.27 sec > 0.7 sec, use the UBC [Sec. 1630.5], Formula 30-14.

$$F_t = 0.07TV$$
$$= (0.07)(1.27 \text{ sec})(90,000 \text{ lbf})$$
$$= 8000 \text{ lbf}$$

Check that $F_t \le 0.25V$.

$$0.25V = (0.25)(90,000 \text{ lbf})$$
$$= 22,500 \text{ lbf}$$

Since 8000 lbf < 22,500 lbf, the value of 8000 lbf is used for $F_t$.

**13. Answer A**
Per UBC [Sec. 1633.2.9, Item 2], the force $F_{Px}$ should not be less than $0.5C_aIw_{Px}$.

From UBC Table 16-Q, the value of seismic coefficient $C_a$ for the soil profile type $S_B$ and $Z = 0.3$ is 0.30.

$$0.5C_aIw_{Px} = (0.5)(0.30)(1.0)w_{Px}$$
$$= 0.15w_{Px}$$

**14. Answer C**

*SI solution*

From the UBC [Sec. 1630.2.2, Item1], $C_t$ for all these special steel moment-resisting buildings is 0.0853. From UBC Formula 30-8,

$$T = C_t(h_n)^{3/4}$$
$$T_I = (0.0853)(9.1 \text{ m})^{3/4} = 0.45 \text{ s}$$
$$T_{II} = (0.0853)(18.2 \text{ m})^{3/4} = 0.75 \text{ s}$$
$$T_{III} = (0.0853)(27.3 \text{ m})^{3/4} = 1.00 \text{ s}$$

$T_{III}$ has the largest natural period.

*Customary U.S. solution*

From the UBC [Sec. 1630.2.2, Item 1], $C_t$ for all these special steel moment-resisting buildings is 0.035. From UBC Formula 30-8,

$$T = C_t(h_n)^{3/4}$$
$$T_I = (0.035)(30 \text{ ft})^{3/4} = 0.45 \text{ sec}$$
$$T_{II} = (0.035)(60 \text{ ft})^{3/4} = 0.75 \text{ sec}$$
$$T_{III} = (0.035)(90 \text{ ft})^{3/4} = 1.00 \text{ sec}$$

$T_{III}$ has the largest natural period.

**15. Answer D**

*SI solution*

For buildings I, II, and III, the dense soil measured shear wave velocity, $\overline{\gamma}_s$, is 760 m/s and the rock measured shear velocity, $\overline{\gamma}_s$, is 1500 m/s.

Per the UBC [Sec. 1636.2.1], to obtain average shear wave velocity for each building, UBC Formula 36-1 should be used.

$$\overline{\gamma}_s = \frac{\sum\limits_{i=1}^{n} d_i}{\sum\limits_{i=1}^{n} \dfrac{d_i}{\gamma_{si}}}$$

For building I,

$$\overline{\gamma}_s = \frac{6.1 \text{ m} + 42.7 \text{ m}}{\dfrac{6.1 \text{ m}}{760 \dfrac{\text{m}}{\text{s}}} + \dfrac{42.7 \text{ m}}{1500 \dfrac{\text{m}}{\text{s}}}}$$

$$= 1337.24 \text{ m/s}$$

For building II,

$$\overline{\gamma}_s = \frac{12.2 \text{ m} + 36.6 \text{ m}}{\dfrac{12.2 \text{ m}}{760 \dfrac{\text{m}}{\text{s}}} + \dfrac{36.6 \text{ m}}{1500 \dfrac{\text{m}}{\text{s}}}}$$

$$= 1206.35 \text{ m/s}$$

For building III,

$$\overline{\gamma}_s = \frac{18.3 \text{ m} + 30.5 \text{ m}}{\dfrac{18.3 \text{ m}}{760 \dfrac{\text{m}}{\text{s}}} + \dfrac{30.5 \text{ m}}{1500 \dfrac{\text{m}}{\text{s}}}}$$

$$= 1098.80 \text{ m/s}$$

Based on the UBC [Sec. 1636.2] and UBC Table 16-J, soil profile type $S_B$ can be classified for buildings I, II, and III because the shear wave velocity of soil profile type $S_B$ is between 760 m/s and 1500 m/s.

Based on UBC Table 16-U, for buildings I, II, and III, the seismic source type is C because the maximum moment magnitude, $M$, is less than 6.5 and slip rate is equal to 2 mm/yr. It is important to note that the UBC requires both maximum moment magnitude and slip rate conditions be satisfied concurrently when determining the seismic source type.

Per UBC Table 16-R, the seismic coefficient $C_v$ with soil profile $S_B$ in seismic zone 4 is $0.40N_v$. From UBC Table 16-T, with seismic source type C and the building being 15 km away from known seismic source, near-source factor $N_v$ is 1.0 for buildings I, II, and III.

Therefore,
$$C_v = (0.40)(1.0)$$
$$= 0.4$$

*Customary U.S. solution*

For buildings I, II, and III, the dense soil measured shear wave velocity, $\overline{\gamma}_s$, is 2500 ft/sec and the rock measured shear velocity, $\overline{\gamma}_s$, is 5000 ft/sec.

Per the UBC [Sec. 1636.2.1], to obtain average shear wave velocity for each building, UBC Formula 36-1 should be used.

$$\overline{\gamma}_s = \frac{\sum\limits_{i=1}^{n} d_i}{\sum\limits_{i=1}^{n} \dfrac{d_i}{\gamma_{si}}}$$

For building I,

$$\overline{\gamma}_s = \frac{20 \text{ ft} + 140 \text{ ft}}{\dfrac{20 \text{ ft}}{2500 \ \dfrac{\text{ft}}{\text{sec}}} + \dfrac{140 \text{ lbf}}{5000 \ \dfrac{\text{ft}}{\text{sec}}}}$$

$$= 4444.44 \text{ ft/sec}$$

For building II,

$$\overline{\gamma}_s = \frac{40 \text{ ft} + 120 \text{ ft}}{\dfrac{40 \text{ ft}}{2500 \ \dfrac{\text{ft}}{\text{sec}}} + \dfrac{120 \text{ ft}}{5000 \ \dfrac{\text{ft}}{\text{sec}}}}$$

$$= 4000.00 \text{ ft/sec}$$

For building III,

$$\overline{\gamma}_s = \frac{60 \text{ ft} + 100 \text{ ft}}{\dfrac{60 \text{ ft}}{2500 \ \dfrac{\text{ft}}{\text{sec}}} + \dfrac{100 \text{ ft}}{5000 \ \dfrac{\text{ft}}{\text{sec}}}}$$

$$= 3636.36 \text{ ft/sec}$$

Based on the UBC [Sec. 1636.2] and UBC Table 16-J, soil profile type $S_B$ can be classified for buildings I, II, and III because the shear wave velocity of soil profile type $S_B$ is between 2500 ft/sec and 5000 ft/sec.

Based on UBC Table 16-U, for buildings I, II, and III, the seismic source type is C because the maximum moment magnitude, $M$, is less than 6.5, and the slip rate is equal to 2 mm/yr. It is important to note that the UBC requires both maximum moment magnitude and slip rate conditions to be satisfied concurrently when determining the seismic source type.

Per UBC Table 16-R, the seismic coefficient $C_v$ with soil profile $S_B$ in seismic zone 4 is $0.40N_v$. From UBC Table 16-T, with seismic source type C and the building 9.32 mi away from known seismic source, near-source factor $N_v$ is 1.0 for buildings I, II, and III.

Therefore,

$$C_v = (0.40)(1.0)$$
$$= 0.4$$

**16. Answer B**

*SI solution*

$$h_n = (3)(4.6 \text{ m}) = 13.8 \text{ m}$$

From the UBC [Sec. 1630.2.2, Item 1], $C_t$ is 0.0853. Use UBC Formula 30-8.

$$T = C_t(h_n)^{3/4}$$
$$= (0.0853)(13.8 \text{ m})^{3/4}$$
$$= 0.6 \text{ s}$$

Since 0.6 s < 0.7 s, per the UBC [Sec. 1630.5], $F_t = 0$.

| level | $h_x$ | $w_x$ | $h_x w_x$ |
|---|---|---|---|
| 3 (roof) | 13.8 m | 1334.46 kN | 18 416 kN·m |
| | 9.2 m | 1779.28 kN | 16 369 kN·m |
| | 4.6 m | 2224.10 kN | 10 231 kN·m |
| | | | $\sum$ 45 016 kN·m |

From UBC Formula 30-15,

$$F_x = \frac{(V - F_t)w_x h_x}{\sum\limits_{i=1}^{n} w_i h_i}$$

Determine the base shear at the roof level.

$$F_3 = \frac{(444\,820 \text{ N} - 0 \text{ N})(18\,416 \text{ kN·m})}{45\,016 \text{ kN·m}}$$
$$= 181\,975 \text{ N}$$

*Customary U.S. solution*

$$h_n = (3)(15 \text{ ft}) = 45 \text{ ft}$$

From the UBC [Sec. 1630.2.2, Item 1], $C_t$ is 0.035. Use UBC Formula 30-8.

$$T = C_t(h_n)^{3/4}$$
$$= (0.035)(45 \text{ ft})^{3/4}$$
$$= 0.6 \text{ sec}$$

Since 0.6 sec < 0.7 sec, per the UBC [Sec. 1630.5],

$$F_t = 0$$

| level | $h_x$ | $w_x$ | $h_x w_x$ |
|---|---|---|---|
| 3 (roof) | 45 ft | 300 k | 13,500 ft-k |
| 2 | 30 ft | 400 k | 12,000 ft-k |
| 1 | 15 ft | 500 k | 7500 ft-k |
| | | | $\sum$ 33,000 ft-k |

From UBC Formula 30-15,

$$F_x = \frac{(V - F_t)w_x h_x}{\sum\limits_{i=1}^{n} w_i h_i}$$

Determine the base shear at the roof level.

$$F_3 = \frac{(100 \text{ k} - 0 \text{ k})(13{,}500 \text{ ft-k})}{33{,}000 \text{ ft-k}}$$
$$= 40.9 \text{ k}$$
$$= 41{,}000 \text{ lbf}$$

**17. Answer C**

*SI solution*

The overturning moment at the base is simply the sum of the overturning moments due to the seismic forces ($F_t$ and $F_x$) at each level.

$$M_{\text{overturning}} = (111 \text{ kN})(12 \text{ m}) + (102 \text{ kN})(9 \text{ m})$$
$$+ (80 \text{ kN})(6 \text{ m}) + (67 \text{ kN})(3 \text{ m})$$
$$= 2931 \text{ kN·m}$$

*Customary U.S. solution*

The overturning moment at the base is simply the sum of the overturning moments due to the seismic forces ($F_t$ and $F_x$) at each level.

$$M_{\text{overturning}} = (25 \text{ k})(40 \text{ ft}) + (23 \text{ k})(30 \text{ ft})$$
$$+ (18 \text{ k})(20 \text{ ft}) + (5 \text{ k})(10 \text{ ft})$$
$$= 2200 \text{ ft-k}$$

**18. Answer B**

Contrary to rigid diaphragms that distribute lateral forces in proportion to the rigidities of vertical resisting elements, flexible diaphragms (i.e., wood or light steel) distribute lateral forces to vertical resisting elements in proportion to the tributary area of the elements.

**19. Answer A**

Flexible diaphragms are relatively thin structural elements that are anchored to the vertical resisting elements. They lack bending strength and depend on the stiffness of perpendicular walls to limit overall

diaphragm deflection. Flexible diagrams are incapable of distributing torsional moments to vertical resisting elements.

**20. Answer D**

Steel decks with zonolite are structurally relatively thin and are considered flexible diaphragms.

**21. Answer A**

Rigid diaphragms distribute lateral forces in proportion to the rigidities of vertical resisting elements. They do not distribute lateral forces on a tributary load basis as do flexible diaphragms.

**22. Answer D**

Rigid diaphragms, such as concrete slabs and concrete on metal deck floor systems, do not bend due to lateral loads. They distribute lateral story shears to vertical resisting elements. As opposed to flexible diaphragms, rigid diaphragms transmit torsion to the vertical resisting elements.

**23. Answer D**

Metal decks with concrete fill are considered rigid diaphragms because they are structurally relatively thick and heavy and will not change shape under lateral loads.

**24. Answer C**

UBC provisions limit diaphragm deflection by specifying maximum diaphragm dimension ratios in Table 23-II-G. One way to further minimize diaphragm deflection is to decrease the nail spacing along the edges of the diaphragm. The nail spacing determines the shear resistance across the diaphragm when it is subjected to lateral loads.

**25. Answer C**

Following are a few definitions that are required in order to answer this question.

- The *center of rigidity* is the point through which the resultant of the resistance to the applied lateral force acts. It is called the center of rigidity because the individual elements resist the lateral force in proportion to their rigidities.

- The *center of mass* is the point through which the applied lateral force acts.

- *Eccentricity* is the distance between the center of rigidity and the center of mass (measured perpendicular to the direction of the lateral load).

- *Accidental eccentricity* is equal to 5% of the building dimension that is perpendicular to the direction of the lateral load.
- The *torsional moment* equals the applied lateral force multiplied by the sum of the eccentricity and accidental eccentricity. The torsional moment is resisted by individual walls and columns in proportion to the distance from the center of rigidity.

Therefore, for elements that are located far from the center of rigidity, torsional effects are critical.

## 26. Answer D

Torsional shear stress occurs whenever centers of mass and rigidity do not coincide. The implication here is that diaphragms are rigid. Rigid diaphragms (as opposed to flexible diaphragms) transfer torsion to the vertical resisting elements. Resisting elements consist of columns, shear walls (parallel and perpendicular), braced frames, moment-resisting frames, and connections.

## 27. Answer C

Based on the UBC [Sec. 1630.7], the torsional design moment should be the moment resulting from eccentricity and accidental torsion.

## 28. Answer C

*SI solution*

Each of the two parallel walls carries half of the applied load in shear. The total load on the diaphragm is

$$V_{\text{diaphragm}} = w(24.4 \text{ m})$$

The load on each parallel wall is

$$V_{\text{wall}} = \frac{(24.4 \text{ m})w}{2}$$
$$= (12.2 \text{ m})w$$

The shear stress per meter of wall is

$$\vartheta = \frac{V_{\text{wall}}}{b} = \frac{(12.2 \text{ m})w}{12.2 \text{ m}}$$
$$= w$$

*Customary U.S. solution*

Each of the two parallel walls carries half of the applied load in shear. The total load on the diaphragm is

$F_{DIAPHRARM}:$ $V_{\text{diaphragm}} = w(80 \text{ ft})$

The load on each parallel wall is

$$V_{\text{wall}} = \frac{(80 \text{ ft})w}{2}$$
$$= (40 \text{ ft})w$$

The shear stress per foot of wall is

$$\vartheta = \frac{V_{\text{wall}}}{b} = \frac{(40 \text{ ft})w}{40 \text{ ft}}$$
$$= w$$

## 29. Answer B

*SI solution*

$$\vartheta = \frac{V}{b}$$
$$V = \vartheta b$$
$$V = \frac{wL}{2}$$

Thus,

$$w = \frac{2V}{L}$$
$$= \frac{2\vartheta b}{L}$$
$$= \frac{(2)\left(3648 \dfrac{\text{N}}{\text{m}}\right)(24.4 \text{ m})}{54.9 \text{ m}}$$
$$= 3243 \text{ N/m}$$

*Customary U.S. solution*

DIAPHLASM
SHEAR CAPACITY $\left(\vartheta = \frac{V}{b}\right)$
$$V = \vartheta b$$
$$V = \frac{wL}{2}$$

Thus,

$$w = \frac{2V}{L}$$
$$= \frac{2\vartheta b}{L}$$
$$= \frac{(2)\left(250 \dfrac{\text{lbf}}{\text{ft}}\right)(80 \text{ ft})}{(180 \text{ ft})}$$
$$= 222 \text{ lbf/ft}$$

**30. Answer B**

*SI solution*

Find $V$, the total force carried by each wall, and then determine the length needed to keep the shear stress at 5838 N/m.

$$V = \frac{wL}{2} = \frac{\left(3648 \frac{N}{m}\right)(24.4 \text{ m})}{2}$$

$$= 44\,506 \text{ N}$$

$$\vartheta = \frac{V}{b}$$

$$b = \frac{V}{\vartheta}$$

$$= \frac{44\,506 \text{ N}}{5838 \frac{N}{m}}$$

$$= 7.6 \text{ m}$$

*Customary U.S. solution*

Find $V$, the total force carried by each wall, and then determine the length needed to keep the shear stress at 400 lbf/ft.

$$V = \frac{wL}{2} = \frac{\left(250 \frac{\text{lbf}}{\text{ft}}\right)(80 \text{ ft})}{2}$$

$$= 10,000 \text{ lbf}$$

$$\vartheta = \frac{V}{b}$$

$$b = \frac{V}{\vartheta}$$

$$= \frac{10,000 \text{ lbf}}{400 \frac{\text{lbf}}{\text{ft}}}$$

$$= 25 \text{ ft}$$

**31. Answer D**

*SI solution*

Write the equation for unit shear stress in the resisting walls for both the N-S and E-W loading conditions. In each direction, the shear stress is dependent on the length of the E-W walls.

N-S loading:

$$\vartheta = \frac{\left(2189 \frac{N}{m}\right)(30.5 \text{ m})}{2} = 33\,382 \text{ N}/b \quad \text{[Eq.1]}$$

E-W loading:

$$\vartheta = \left(\frac{\left(4378 \frac{N}{m}\right)b}{2}\right)\left(\frac{1}{30.5 \text{ m}}\right) = 71.8b \quad \text{[Eq.2]}$$

Set Eq. 1 equal to Eq. 2 and solve for $b$.

$$\frac{33\,382}{b} = 71.8b$$

$$71.8b^2 = 33\,382$$

$$b = 21.6 \text{ m}$$

*Customary U.S. solution*

Write the equations for unit shear stress in the resisting walls for both the N-S and E-W loading conditions. In each direction, the shear stress is dependent on the length of the E-W walls.

N-S loading:

$$\vartheta = \frac{\left(150 \frac{\text{lbf}}{\text{ft}}\right)(100 \text{ ft})}{2} = 7500 \text{ lbf}/b \quad \text{[Eq. 1]}$$

E-W loading

$$\vartheta = \left(\frac{\left(300 \frac{\text{lbf}}{\text{ft}}\right)b}{2}\right)\left(\frac{1}{100 \text{ ft}}\right) = 1.5b \quad \text{[Eq. 2]}$$

Set Eq. 1 equal to Eq. 2 and solve for $b$.

$$\frac{7500}{b} = 1.5b$$

$$1.5b^2 = 7500$$

$$b = 70.71 \text{ ft}$$

32. Answer C

*SI solution*

$$\vartheta = \frac{V}{b}$$
$$V = \vartheta b$$
$$= \left(7297 \ \frac{\text{N}}{\text{m}}\right)(15.2 \ \text{m})$$
$$= 110\,914 \ \text{N}$$

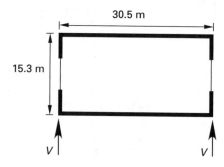

$$V = \frac{wL}{2}$$
$$w = \frac{2V}{L}$$
$$= (2)\left(\frac{110\,914 \ \text{N}}{30.5 \ \text{m}}\right)$$
$$= 7273 \ \text{N/m}$$

*Customary U.S. solution*

$$\vartheta = \frac{V}{b}$$
$$V = \vartheta b$$
$$= \left(500 \ \frac{\text{lbf}}{\text{ft}}\right)(50 \ \text{ft})$$
$$= 25,000 \ \text{lbf}$$

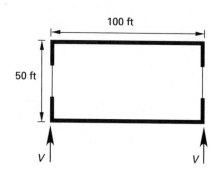

$$V = \frac{wL}{2}$$
$$w = \frac{2V}{L}$$
$$= (2)\left(\frac{25,000 \ \text{lbf}}{100 \ \text{ft}}\right)$$
$$= 500 \ \text{lbf/ft}$$

33. Answer C

*SI solution*

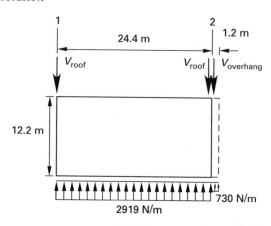

The tributary area to line 1 includes everything to the plane halfway between lines 1 and 2. The seismic load from the roof overhang is tributary to line 2.

$$V = \frac{\left(2919 \ \dfrac{\text{N}}{\text{m}}\right)(24.4 \ \text{m})}{2}$$
$$= \frac{71\,223.6 \ \text{N}}{2}$$
$$= 35\,612 \ \text{N}$$
$$\approx 35\,600 \ \text{N}$$

*Customary U.S. solution*

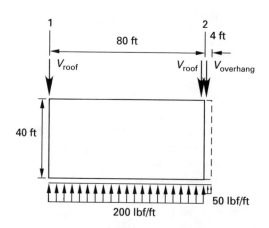

The tributary area to line 1 includes everything to the plane halfway between lines 1 and 2. The seismic load from the roof overhang is tributary to line 2.

$$V = \frac{\left(200 \, \frac{\text{lbf}}{\text{ft}}\right)(80 \text{ ft})}{2}$$
$$= \frac{16{,}000 \text{ lbf}}{2}$$
$$= 8000 \text{ lbf}$$

**34. Answer D**

*SI solution*

The tributary area to line 2 includes everything to the plane halfway between lines 1 and 2, plus the roof overhang. As determined in Prob. 33, $V_{\text{roof at line 2}} = 35\,586$ N. Therefore,

$$V_{\text{overhang at line 2}} = \left(730 \, \frac{\text{N}}{\text{m}}\right)(1.2 \text{ m})$$
$$= 876 \text{ N}$$
$$V_{\text{total}} = 35\,600 \text{ N} + 876 \text{ N}$$
$$= 36\,476 \text{ N}$$
$$\approx 36\,480 \text{ N}$$

*Customary U.S. solution*

The tributary area to line 2 includes everything to the plane halfway between lines 1 and 2, plus the roof overhang. As determined in Prob. 33, $V_{\text{roof at line 2}} = 8000$ lbf. Therefore,

$$V_{\text{overhang at line 2}} = \left(50 \, \frac{\text{lbf}}{\text{ft}}\right)(4 \text{ ft})$$
$$= 200 \text{ lbf}$$
$$V_{\text{total}} = 8000 \text{ lbf} + 200 \text{ lbf}$$
$$= 8200 \text{ lbf}$$

**35. Answer D**

*SI solution*

Diaphragm force on the parallel walls:

This is shear force along resisting lines 1 and 2.

$$V = \frac{wL}{2} = \frac{\left(2919 \, \frac{\text{N}}{\text{m}}\right)(24.4 \text{ m})}{2}$$
$$= 35\,612 \text{ N}$$

Shear wall:

The shear stress in the parallel walls is due to the diaphragm force plus the inertia force of the parallel wall weights. The base shear equation is

$$V = \left(\frac{C_v I}{RT}\right)W$$
$$= 0.1375W$$
$$L_{\text{shear walls,line 2}} = 3 \text{ m} + 6.1 \text{ m}$$
$$= 9.1 \text{ m}$$
$$W_{\text{shear wall,line 2}} = \frac{(9.1 \text{ m})\left(766 \, \frac{\text{N}}{\text{m}^2}\right)(4.3 \text{ m})}{2}$$
$$= 14\,987 \text{ N}$$
$$V = (0.1375)(14\,987 \text{ N})$$
$$= 2061 \text{ N}$$
$$V = V_{\text{roof}} + V_{\text{wall}}$$
$$= 35\,612 \text{ N} + 2061 \text{ N}$$
$$= 37\,673 \text{ N}$$

At line 2 for 6.1 m shear wall:

$$\vartheta = \frac{V}{b}$$
$$= \frac{37\,673 \text{ N}}{3 \text{ m} + 6.1 \text{ m}}$$
$$= 4140 \text{ N/m}$$

*Customary U.S. solution*

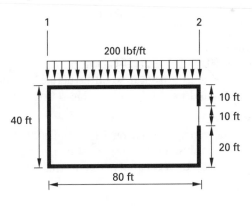

Diaphragm force on the parallel walls:

This is shear force along resisting lines 1 and 2.

$$V = \frac{wL}{2} = \frac{\left(200\ \dfrac{\text{lbf}}{\text{ft}}\right)(80\ \text{ft})}{2}$$
$$= 8000\ \text{lbf}$$

Shear wall:

The shear stress in the parallel walls is due to the diaphragm force plus the inertia force of the parallel wall weights. The base shear equation is

$$V = \left(\frac{C_v I}{RT}\right)W$$
$$= 0.1375W$$
$$L_{\text{shear walls,line 2}} = 10\ \text{ft} + 20\ \text{ft}$$
$$= 30\ \text{ft}$$
$$W_{\text{shear wall,line 2}} = \frac{(30\ \text{ft})\left(16\ \dfrac{\text{lbf}}{\text{ft}^2}\right)(14\ \text{ft})}{2}$$
$$= 3360\ \text{lbf}$$
$$V = (0.1375)(3360\ \text{lbf})$$
$$= 460\ \text{lbf}$$
$$V = V_{\text{roof}} + V_{\text{wall}}$$
$$= 8000\ \text{lbf} + 460\ \text{lbf}$$
$$= 8460\ \text{lbf}$$

At line 2 for 20 ft shear wall:

$$\vartheta = \frac{V}{b}$$
$$= \frac{8460\ \text{lbf}}{10\ \text{ft} + 20\ \text{ft}}$$
$$= 282\ \text{lbf/ft}$$

**36. Answer D**
Based on UBC Table 23-II-G, the maximum span-to-width ratio for horizontal wood structural panel diaphragms should be 4:1 for all seismic zones.

**37. Answer A**
For seismic zone 4, based on UBC Table 23-II-G, the maximum diaphragm dimension ratio (i.e., the height-to-width ratio) for vertical diaphragms (nailed all edges) should be 2:1.

**38. Answer C**

*SI solution*

From UBC Table 16-I, $Z$ is 0.4. From UBC Table 16-N, $R$ is 5.5. From UBC Table 16-Q, seismic coefficient $C_a$ for soil profile type $S_C$ is

$$C_a = 0.4 N_a$$
$$= (0.4)(1.0)$$
$$= 0.4$$

Based on the UBC [Sec. 1630.2.3], use UBC Formula 30-11.

$$V = \left(\frac{3.0 C_a}{R}\right)W$$
$$= \left(\frac{(3.0)(0.4)}{5.5}\right)W$$
$$= 0.2182W$$
$$W_{\text{roof}} = (4)(7.6\ \text{m})(7.6\ \text{m})\left(958\ \frac{\text{N}}{\text{m}^2}\right)$$
$$= 221\,336\ \text{N}$$
$$W_{\text{walls}} = (5\ \text{walls})\left(\frac{4.9\ \text{m}}{2}\right)(7.6\ \text{m})\left(1197\ \frac{\text{N}}{\text{m}^2}\right)$$
$$= 111\,441\ \text{N}$$
$$W_{\text{total}} = 221\,336\ \text{N} + 111\,441\ \text{N}$$
$$= 332\,777\ \text{N}$$
$$V = (0.2182)(332\,777\ \text{N})$$
$$= 72\,612\ \text{N}$$
$$V_{\text{shear at line 1}} = \frac{72\,612\ \text{N}}{2}$$
$$= 36\,306\ \text{N}$$

*Customary U.S. solution*

From UBC Table 16-I, $Z$ is 0.4. From UBC Table 16-N, $R$ is 5.5. From UBC Table 16-Q, seismic coefficient $C_a$ for soil profile type $S_C$ is

$$C_a = 0.4 N_a$$
$$= (0.4)(1.0)$$
$$= 0.4$$

Based on the UBC [Sec. 1630.2.3], use UBC Formula 30-11.

$$V = \left(\frac{3.0C_a}{R}\right)W$$

$$= \left(\frac{(3.0)(0.4)}{5.5}\right)W$$

$$= 0.2182W$$

$$W_{\text{roof}} = (4)(25 \text{ ft})(25 \text{ ft})\left(20 \frac{\text{lbf}}{\text{ft}^2}\right)$$

$$= 50{,}000 \text{ lbf}$$

$$W_{\text{walls}} = (5 \text{ walls})\left(\frac{16 \text{ ft}}{2}\right)(25 \text{ ft})\left(25 \frac{\text{lbf}}{\text{ft}^2}\right)$$

$$= 25{,}000 \text{ lbf}$$

$$W_{\text{total}} = 50{,}000 \text{ lbf} + 25{,}000 \text{ lbf}$$

$$= 75{,}000 \text{ lbf}$$

$$V = (0.2182)(75{,}000 \text{ lbf})$$

$$= 16{,}365 \text{ lbf}$$

$$V_{\text{shear at line 1}} = \frac{16{,}365 \text{ lbf}}{2}$$

$$\approx 8180 \text{ lbf}$$

### 39.  Answer A

*SI solution*

For wood structural shear wall panels that are nailed at their edges, per the UBC [Chap. 23, Table 23-II-G], maximum diaphragm dimension ratios for vertical diaphragms in seismic zone 4 should be 2:1. Therefore, for all panels, check the height-to-width ratios.

For panel I,

$$\frac{3.7}{4.6} = 0.8 \quad [\text{less than 2}]$$

For panel II,

$$\frac{3.7}{0.9} = 4.1 \quad [\text{greater than 2}]$$

For panel III,

$$\frac{3.7}{3.0} = 1.2 \quad [\text{less than 2}]$$

For panel IV,

$$\frac{3.7}{1.5} = 2.4 \quad [\text{greater than 2}]$$

Panels I and III are in compliance with the UBC Maximum Diaphragm Dimension Ratios requirements.

*Customary U.S. solution*

For wood structural shear wall panels that are nailed at their edges, per the UBC [Chap. 23, Table 23-II-G], maximum diaphragm dimension ratios for a vertical diaphragm in seismic zone 4 should be 2:1. Therefore, for all panels, check the height-to-width ratios.

For panel I,

$$\frac{12}{15} = 0.8 \quad [\text{less than 2}]$$

For panel II,

$$\frac{12}{3} = 4.0 \quad [\text{greater than 2}]$$

For panel III,

$$\frac{12}{10} = 1.2 \quad [\text{less than 2}]$$

For panel IV,

$$\frac{12}{5} = 2.4 \quad [\text{greater than 2}]$$

Panels I and III are in compliance with the UBC Maximum Diaphragm Dimension Ratios requirements.

### 40.  Answer C

*SI solution*

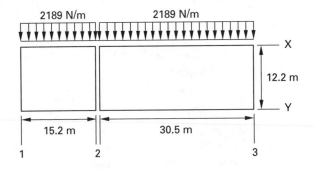

The shear wall along line 3 carries half of the diaphragm shear between lines 2 and 3.

$$V = \frac{wL}{2}$$

$$= \frac{\left(2189 \frac{\text{N}}{\text{m}}\right)(30.5 \text{ m})}{2}$$

$$= 33\,382 \text{ N}$$

*Customary U.S. solution*

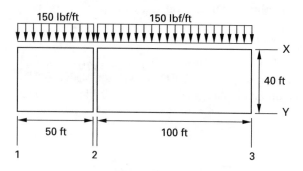

The shear wall along line 3 carries half of the diaphragm shear between lines 2 and 3.

$$V = \frac{wL}{2}$$
$$= \frac{\left(150 \, \dfrac{\text{lbf}}{\text{ft}}\right)(100 \text{ ft})}{2}$$
$$= 7500 \text{ lbf}$$

**41. Answer B**

*SI solution*

The maximum diaphragm chord force occurs at the midpoint between lines 1 and 2. It is calculated as the bending moment of a simple beam under a distributed load, per unit depth of diaphragm.

$$C = \frac{M}{b} = \frac{wL^2}{8b}$$
$$= \frac{\left(2189 \, \dfrac{\text{N}}{\text{m}}\right)(15.2 \text{ m})^2}{(8)(12.2 \text{ m})}$$
$$= 5182 \text{ N}$$

*Customary U.S. solution*

The maximum diaphragm chord force occurs at the midpoint between lines 1 and 2. It is calculated as the bending moment of a simple beam under a distributed load, per unit depth of diaphragm.

$$C = \frac{M}{b} = \frac{wL^2}{8b}$$
$$= \frac{\left(150 \, \dfrac{\text{lbf}}{\text{ft}}\right)(50 \text{ ft})^2}{(8)(40 \text{ ft})}$$
$$= 1172 \text{ lbf}$$

**42. Answer A**

Chords are boundary members of a diaphragm that are perpendicular to the direction of the lateral load. Chords are designed to carry moment and provide all the resistance to the flexural stresses. Chord forces at the chord ends are zero.

**43. Answer B**

*SI solution*

$$b = 3.0 \text{ m} + 6.1 \text{ m} + 1.5 \text{ m}$$
$$= 10.6 \text{ m}$$
$$\vartheta = \frac{V}{b}$$
$$= \frac{40\,034 \text{ N}}{10.6 \text{ m}}$$
$$= 3776.8 \text{ N/m}$$
$$V = \left(3776.8 \, \frac{\text{N}}{\text{m}}\right)(3.0 \text{ m})$$
$$= 11\,330 \text{ N}$$

*Customary U.S. solution*

$$b = 10 \text{ ft} + 20 \text{ ft} + 5 \text{ ft}$$
$$= 35 \text{ ft}$$
$$\vartheta = \frac{V}{b}$$
$$= \frac{9000 \text{ lbf}}{35 \text{ ft}}$$
$$= 257.1 \text{ lbf/ft}$$
$$V = \left(257 \, \frac{\text{lbf}}{\text{ft}}\right)(10 \text{ ft})$$
$$= 2570 \text{ lbf}$$

**44. Answer B**

At points of discontinuity in the plan, or where the vertical resisting elements are not provided because of windows or doors, the drag struts collect and drag the horizontal diaphragm shear to the supporting vertical resisting elements.

**45. Answer C**

Struts collect diaphragm load and carry it to a shear wall. Struts carry tension and compression depending on the direction of seismic loading.

**46. Answer A**

The strut force changes along the length of the drag strut. The maximum force occurs at the point where the drag strut attaches to the parallel wall. For the north shear wall, considering the direction of seismic loading shown, this location is at point X.

**47. Answer D**

The strut force changes along the length of the drag strut. The maximum force occurs at the point where the drag strut attaches to the parallel wall. Points X and Z frame into parallel walls. Therefore, the maximum drag strut forces occur at both locations.

**48. Answer C**

Struts carry tension and compression, depending on the direction of seismic loading. The outside ends of the walls can be considered to experience no strut forces. The critical loading conditions depend on the greater magnitude of the strut and chord forces at the corresponding points along a wall.

**49. Answer C**

Chords are elements, such as walls or reinforcement at the diaphragm boundaries that are perpendicular to the direction of the applied seismic loading. They resist moments and are regarded as tension and compression members.

**50. Answer D**

Chords are elements at the borders of the diaphragm along the walls perpendicular to the direction of the applied seismic loading. Any continuous material that resists all tensile and compressive forces can be used. For example, masonry walls with adequate tensile reinforcement, the double top-plate in wood stud walls, and steel are suitable.

**51. Answer D**

Chord force, $C$, is calculated as the bending moment of a simple beam due to a distributed load, per unit depth of the diaphragm.

$$C = \frac{M}{b} \quad \text{[Eq. 1]}$$

The bending moment of a simple beam is found in any elementary mechanics textbook.

$$M = \frac{wL^2}{8} \quad \text{[Eq. 2]}$$

Insert Eq. 2 into Eq. 1 and solve for the chord force.

$$C = \frac{M}{b} = \frac{wL^2}{8b} \quad \text{[Eq. 3]}$$

The distributed load is the diaphragm force divided by the span.

$$w = \frac{F}{L} \quad \text{[Eq. 4]}$$

Insert Eq. 4 into Eq. 3 and solve for the chord force.

$$C = \frac{wL^2}{8b} = \frac{\left(\dfrac{F}{L}\right)L^2}{8b}$$
$$= \frac{FL}{8b}$$

All three formulas for the chord force are correct.

**52. Answer A**

The chords at lines 1 and 2 resist bending due to the diaphragm loads, when the loading is in the direction shown. At both lines, the maximum chord forces are equal and occur at mid-span where the maximum moment develops.

**53. Answer A**

*SI solution*

The diaphragm shear is resisted by the parallel walls. The shear force on each wall is

$$V = \vartheta b$$
$$= \left(2919 \ \frac{\text{N}}{\text{m}}\right)(15.2 \ \text{m})$$
$$= 44\,369 \ \text{N}$$

Each parallel wall resists half of the diaphragm load. The total distributed diaphragm load is

$$w = \frac{2V}{L}$$
$$= \frac{(2)(44\,369 \ \text{N})}{45.7 \ \text{N}}$$
$$= 1942 \ \text{N/m}$$

The maximum chord force is calculated as the maximum bending stress due to the distributed diaphragm load.

$$C = \frac{M}{b}$$

$$M_{\max} = \frac{wL^2}{8}$$

$$C_{\max} = \frac{wL^2}{8b}$$

$$= \frac{\left(1942 \ \frac{\text{N}}{\text{m}}\right)(45.7 \ \text{m})^2}{(8)(15.2 \ \text{m})}$$

$$= 33\,354 \ \text{N}$$

*Customary U.S. solution*

The diaphragm shear is resisted by the parallel walls. The shear force on each wall is

$$V = \vartheta b$$

$$= \left(200 \ \frac{\text{lbf}}{\text{ft}}\right)(50 \ \text{ft})$$

$$= 10,000 \ \text{lbf}$$

Each parallel wall resists half of the diaphragm load. The total distributed diaphragm load is

$$w = \frac{2V}{L}$$

$$= \frac{(2)(10,000 \ \text{lbf})}{(150 \ \text{ft})}$$

$$= 133.3 \ \text{lbf/ft}$$

The maximum chord force is calculated as the maximum bending stress due to the distributed diaphragm load.

$$C = \frac{M}{b}$$

$$M_{\max} = \frac{wL^2}{8}$$

$$C_{\max} = \frac{wL^2}{8b}$$

$$= \frac{\left(133.3 \ \frac{\text{lbf}}{\text{ft}}\right)(150 \ \text{ft})^2}{(8)(50 \ \text{ft})}$$

$$= 7498 \ \text{lbf}$$

**54. Answer A**

*SI solution*

The chord forces at the chord ends are always zero. This can be demonstrated for the current set of assumptions, but the results can be generalized to any chord member.

The chord force is the resistance to bending under the imposed diaphragm loads. The diaphragm shear at the parallel walls is

$$V = \frac{wL}{2}$$

$$= \frac{\left(3648 \ \frac{\text{N}}{\text{m}}\right)(24.4 \ \text{m})}{2}$$

$$= 44\,506 \ \text{N}$$

The shear stress in the parallel walls is

$$\vartheta = \frac{V}{b}$$

$$= \frac{44\,506 \ \text{N}}{12.2 \ \text{m}}$$

$$= 3648 \ \text{N/m}$$

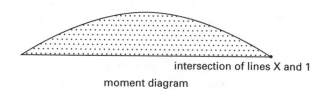

3648 N/m

3648 N/m

shear diagram

intersection of lines X and 1

moment diagram

As shown by the shear and bending moment diagrams, the moments at the ends of the chords are zero. Therefore, the chord forces at the ends are also zero.

$$C = \frac{M}{b}$$

Therefore, $C = 0$.

The force in the chord member at the intersection of lines X and 1 is zero.

*Customary U.S. solution*

The chord forces at the chord ends are always zero. This can be demonstrated for the current set of assumptions, but the results can be generalized to any chord member.

The chord force is the resistance to bending under the imposed diaphragm loads. The diaphragm shear at the parallel walls is

$$V = \frac{wL}{2}$$
$$= \frac{\left(250 \frac{\text{lbf}}{\text{ft}}\right)(80 \text{ ft})}{2}$$
$$= 10,000 \text{ lbf}$$

The shear stress in the parallel walls is

$$\vartheta = \frac{V}{b}$$
$$= \frac{10,000 \frac{\text{lbf}}{\text{ft}}}{40 \text{ ft}}$$
$$= 250 \text{ lbf/ft}$$

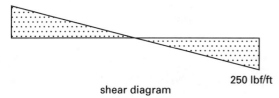

250 lbf/ft

250 lbf/ft

shear diagram

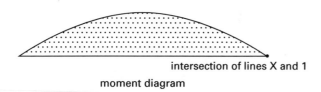

intersection of lines X and 1

moment diagram

As shown by the shear and bending moment diagrams, the moments at the ends of the chords are zero. Therefore, the chord forces at the ends are also zero.

$$C = \frac{M}{b}$$

Therefore, $C = 0$.

The force in the chord member at the intersection of lines X and 1 is zero.

## 55. Answer C

*SI solution*

The chord force is the bending moment per unit depth of the diaphragm. The shear load at the parallel walls is

$$V = \frac{wL}{2}$$
$$= \frac{\left(7297 \frac{\text{N}}{\text{m}}\right)(24.4 \text{ m})}{2}$$
$$= 89\,023 \text{ N}$$

The shear and bending moment diaphragms across the length of the chord are as shown.

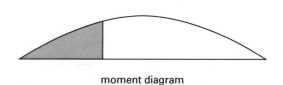

89 023 N

intersection of lines X and 1

9.1 m    3 m    12.2 m    89 023 N

shear diagram

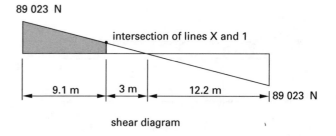

moment diagram

$$\frac{89\,023 \text{ N}}{12.2 \text{ m}} = \frac{V}{3 \text{ m}}$$
$$V = 21\,891 \text{ N} \quad [\text{at lines X and 1}]$$

The moment at line 1 is calculated as the area under the shear diagram.

The chord force is the bending moment divided by the diaphragm depth.

$$M = (21\,891 \text{ N})(9.1 \text{ m}) + \frac{(67\,132 \text{ N})(9.1 \text{ m})}{2}$$
$$= 504\,659 \text{ N·m}$$
$$C = \frac{M}{b}$$
$$= \frac{504\,659 \text{ N·m}}{12.2 \text{ m}}$$
$$= 41\,365 \text{ N}$$

*Customary U.S. solution*

The chord force is the bending moment per unit depth of the diaphragm. The shear load at the parallel walls is

$$V = \frac{wL}{2}$$
$$= \frac{\left(500 \, \frac{\text{lbf}}{\text{ft}}\right)(80 \, \text{ft})}{2}$$
$$= 20{,}000 \, \text{lbf}$$

The shear and bending moment diaphragms across the length of the chord are as shown.

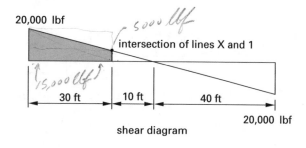

20,000 lbf     5000 llf

intersection of lines X and 1

15,000 llf

| 30 ft | 10 ft | 40 ft |

20,000 lbf

shear diagram

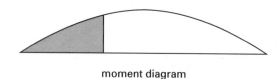

moment diagram

$$\frac{20{,}000 \, \text{lbf}}{40} = \frac{V}{10}$$
$$V = 5000 \, \text{lbf} \quad \text{[at lines X and 1]}$$

The moment at line 1 is calculated as the area under the shear diagram.

The chord force is the bending moment divided by the diaphragm depth.

$$M = (5000 \, \text{lbf})(30 \, \text{ft}) + \frac{(15{,}000 \, \text{lbf})(30 \, \text{ft})}{2}$$
$$= 375{,}000 \, \text{ft-lbf}$$
$$C = \frac{M}{b}$$
$$= \frac{375{,}000 \, \text{ft-lbf}}{40 \, \text{ft}}$$
$$= 9375 \, \text{lbf}$$

**56. Answer A**

The maximum chord force develops at mid-span (at the location of the maximum moment), while the minimum chord force occurs at the chord ends and is zero.

**57. Answer D**

*SI solution*

Assume $x$ equals the distance of an additional wall section measured from the east wall. For the strut force on the south wall, E-W seismic loading applies.

$$V = \frac{wL}{2}$$
$$= \frac{\left(8756 \, \frac{\text{N}}{\text{m}}\right)(15.2 \, \text{m})}{2}$$
$$= 66\,546 \, \text{N}$$
$$\vartheta = \frac{V}{b}$$
$$= \frac{66\,546 \, \text{N}}{30.5 \, \text{m}}$$
$$= 2182 \, \text{N/m}$$

Therefore, the strut force on the south wall is

$$\left(2182 \, \frac{\text{N}}{\text{m}}\right)(x \, \text{m}) = 2182x \, \text{N}$$

For the chord force on the south wall, N-S seismic loading applies. Assume $x$ equals the distance of an additional wall section measured from the east wall.

$$V = \frac{wL}{2}$$
$$= \frac{\left(2919 \, \frac{\text{N}}{\text{m}}\right)(30.5 \, \text{m})}{2}$$
$$= 44\,515 \, \text{N}$$

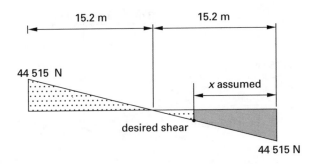

shear diagram

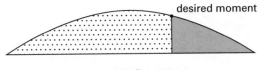

moment diagram

Use the shear and moment diagrams.

The desired shear is

$$V = \left(\frac{44\,515 \text{ N}}{15.2 \text{ m}}\right)(15.2 \text{ m} - x)$$
$$= (2928.6)(15.2 \text{ m} - x)$$

The desired moment is

$$M = \left[(2928.6)(15.2 \text{ m} - x) + (44\,515 \text{ N})\right]\left(\frac{x}{2}\right)$$
$$= 44\,515x - 1464.3x^2$$

The chord force on the south wall is

$$C = \frac{M}{b}$$
$$= \frac{44\,515x - 1464.3x^2}{15.2 \text{ m}}$$
$$= 2928.6x - 96.3x^2$$

Here it is required to find out how long the 10.7 m south wall panel should be in order to have equal magnitude strut and chord forces.

Therefore, the strut force should be set equal to the chord force.

$$2182x = 2928.6x - 96.3x^2$$

Solve for $x$ (shear wall additional length measured from the east wall).

$$x = 7.7 \text{ m}$$
$$\text{south wall}_{\text{extended length}} = 30.5 \text{ m} - 7.7 \text{ m}$$
$$= 22.8 \text{ m}$$

*Customary U.S. solution*

Assume $x$ equals the distance of an additional wall section measured from the east wall. For the strut force on the south wall, E-W seismic loading applies.

$$V = \frac{wL}{2}$$
$$= \frac{\left(600\, \frac{\text{lbf}}{\text{ft}}\right)(50 \text{ ft})}{2}$$
$$= 15,000 \text{ lbf}$$
$$\vartheta = \frac{V}{b}$$
$$= \frac{15,000 \text{ lbf}}{100 \text{ ft}}$$
$$= 150 \text{ lbf/ft}$$

Therefore, the strut force on the south wall is

$$\left(150\, \frac{\text{lbf}}{\text{ft}}\right)(x \text{ ft}) = 150x \text{ lbf}$$

For the chord force on the south wall, N-S seismic loading applies. Assume $x$ equals the distance of an additional wall section measured from the east wall.

$$V = \frac{wL}{2}$$
$$= \frac{\left(200\, \frac{\text{lbf}}{\text{ft}}\right)(100 \text{ ft})}{2}$$
$$= 10,000 \text{ lbf}$$

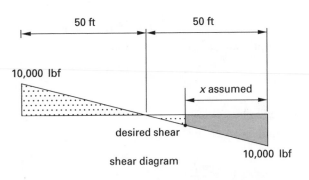

shear diagram

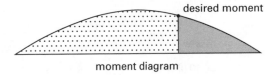

moment diagram

Use the shear and moment diagrams.

The desired shear is

$$V = \left(\frac{10{,}000 \text{ lbf}}{50 \text{ ft}}\right)(50 \text{ ft} - x)$$
$$= (200)(50 \text{ ft} - x)$$

The desired moment is

$$M = [(200)(50 \text{ ft} - x) + (10{,}000 \text{ lbf})]\left(\frac{x}{2}\right)$$
$$= 10{,}000x - 100x^2$$

The chord force on the south wall is

$$C = \frac{M}{b}$$
$$= \frac{10{,}000x - 100x^2}{50 \text{ ft}}$$
$$= 200x - 2x^2$$

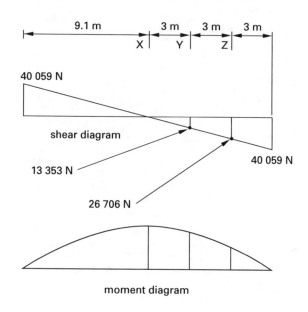

shear diagram

moment diagram

Here it is required to find out how long the 35 ft south wall panel should be in order to have equal magnitude strut and chord forces.

Therefore, the strut force should be set equal to the chord force.

$$150x = 200x - 2x^2$$

Solve for $x$ (shear wall additional length measured from the east wall).

$$x = 25 \text{ ft}$$
$$\text{south wall}_{\text{extended length}} = 100 \text{ ft} - 25 \text{ ft}$$
$$= 75 \text{ ft}$$

58. Answer A

*SI solution*

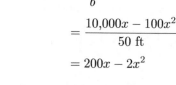

$$V = \frac{wL}{2}$$
$$= \frac{\left(4378 \ \frac{\text{N}}{\text{m}}\right)(18.3 \text{ m})}{2}$$
$$= 40{,}059 \text{ N}$$

Chord force:

$$M_x = \frac{(40\,059 \text{ N})(9.1 \text{ m})}{2}$$
$$= 182\,268 \text{ N·m}$$

$$M_Y = 182\,268 \text{ N·m} - \frac{(13\,353 \text{ N})(3 \text{ m})}{2}$$
$$= 162\,239 \text{ N·m}$$

$$M_Z = 182\,268 \text{ N·m} - \frac{(26\,706 \text{ N})(6.1 \text{ m})}{2}$$
$$= 100\,815 \text{ N·m}$$

$$C = \frac{M}{b}$$
$$C_X = \frac{182\,268 \text{ N·m}}{18.3 \text{ m}}$$
$$= 9960 \text{ N} \approx 10\,000 \text{ N}$$

$$C_Y = \frac{162\,239 \text{ N·m}}{18.3 \text{ m}}$$
$$= 8866 \text{ N} \approx 8900 \text{ N}$$

$$C_Z = \frac{100\,815 \text{ N·m}}{18.3 \text{ m}}$$
$$= 5509 \text{ N} \approx 5500 \text{ N}$$

Strut force:

$$\vartheta = \frac{V}{b}$$

$$\vartheta_{\text{roof}} = \frac{40\,059 \text{ N}}{18.3 \text{ m}}$$
$$= 2189 \text{ N/m}$$

$$\vartheta_{\text{wall}} = \frac{40\,059 \text{ N}}{12.1 \text{ m}}$$
$$= 3311 \text{ N/m}$$

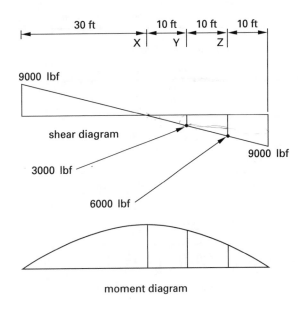

shear diagram

moment diagram

At $X$, the drag strut force is

$$D_X + \left(2189 \ \frac{\text{N}}{\text{m}}\right)(9.1 \text{ m}) - \left(3311 \ \frac{\text{N}}{\text{m}}\right)(3 \text{ m}) = 0$$
$$D_X = -9987 \text{ N} \ \approx 10\,000 \text{ N} \quad [\text{compression}]$$

At $Y$, the drag strut force is

$$D_Y + \left(2189 \ \frac{\text{N}}{\text{m}}\right)(6.1 \text{ m}) - \left(3311 \ \frac{\text{N}}{\text{m}}\right)(3 \text{ m}) = 0$$
$$D_Y = -3420 \text{ N} \ \approx 3400 \text{ N} \quad [\text{compression}]$$

At $Z$, the drag strut force is

$$D_Z + \left(2189 \ \frac{\text{N}}{\text{m}}\right)(3 \text{ m}) - \left(3311 \ \frac{\text{N}}{\text{m}}\right)(3 \text{ m}) = 0$$
$$D_Z = 3366 \text{ N} \ 3400 \text{ N} \quad [\text{tension}]$$

The chord force at $X$ is identical to the drag strut force at $X$.

*Customary U.S. solution*

$$V = \frac{wL}{2}$$
$$= \frac{\left(300 \ \frac{\text{lbf}}{\text{ft}}\right)(60 \text{ ft})}{2}$$
$$= 9000 \text{ lbf}$$

Chord force:

$$M_X = \frac{(9000 \text{ lbf})(30 \text{ ft})}{2}$$
$$= 135,000 \text{ ft-lbf}$$

$$M_Y = 135,000 \text{ ft-lbf} - \frac{(3000 \text{ lbf})(10 \text{ ft})}{2}$$
$$= 120,000 \text{ ft-lbf}$$

$$M_Z = 135,000 \text{ ft-lbf} - \frac{(6000 \text{ lbf})(20 \text{ ft})}{2}$$
$$= 75,000 \text{ ft-lbf}$$

$$C = \frac{M}{b}$$

$$C_X = \frac{135,000 \text{ ft-lbf}}{60 \text{ ft}}$$
$$= 2250 \text{ lbf}$$

$$C_Y = \frac{120,000 \text{ ft-lbf}}{60 \text{ ft}}$$
$$= 2000 \text{ lbf}$$

$$C_Z = \frac{75,000 \text{ ft-lbf}}{60 \text{ ft}}$$
$$= 1250 \text{ lbf}$$

Strut force:

$$\vartheta = \frac{V}{b}$$

$$\vartheta_{\text{roof}} = \frac{9000 \text{ lbf}}{60 \text{ ft}}$$
$$= 150 \text{ lbf/ft}$$

$$\vartheta_{\text{wall}} = \frac{9000 \text{ lbf}}{40 \text{ ft}}$$
$$= 225 \text{ lbf/ft}$$

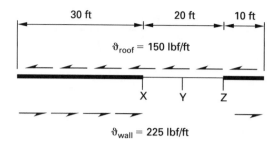

$\vartheta_{roof} = 150$ lbf/ft

X   Y   Z

$\vartheta_{wall} = 225$ lbf/ft

At X, the drag strut force is

$$D_X + \left(150\,\frac{lbf}{ft}\right)(30\text{ ft}) - \left(225\,\frac{lbf}{ft}\right)(10\text{ ft}) = 0$$
$$D_X = -2250\text{ lbf}\quad[\text{compression}]$$

At Y, the drag strut force is

$$D_Y + \left(150\,\frac{lbf}{ft}\right)(20\text{ ft}) - \left(225\,\frac{lbf}{ft}\right)(10\text{ ft}) = 0$$
$$D_Y = -750\text{ lbf}\quad[\text{compression}]$$

At Z, the drag strut force is

$$D_Z + \left(150\,\frac{lbf}{ft}\right)(10\text{ ft}) - \left(225\,\frac{lbf}{ft}\right)(10\text{ ft}) = 0$$
$$D_Z = 750\text{ lbf}\quad[\text{tension}]$$

The chord force at X is identical to the drag strut force at X.

**59. Answer D**
*Drift*, $\Delta$, is the displacement between adjacent stories due to applied forces. The drift can be divided into different components (e.g., shear drift and chord drift) depending on the nature of the loading and the displacement that results. The drift depends on variables, such as the building and story heights, shear load, girder length and depth, column length and height, and frame length.

**60. Answer B**
The *P-$\Delta$ effect* is an additional column bending stress caused by the earthquake. Originally, the vertical loads (live and dead) are constant and concentric with the base of the building. When drift occurs under the horizontal seismic loading, the vertical loads become eccentric with the base. Under these conditions, there are two

types of moment. The primary moment is the bending moment due to bending of the columns under the lateral loads, $(Fh)$. The secondary moment is the additional moment from $P$-$\Delta$ effects, where vertical loads $(P)$ act on the building elements that have been laterally displaced $(\Delta)$ from the base. The $P$-$\Delta$, or secondary moment effects, on shears, axial forces, and moments of frame members should be considered in building design as outlined in the UBC [Sec. 1630.1.3].

**61. Answer A**

*SI solution*

Per the UBC [Sec. 1627], *story drift ratio* is the story drift divided by the story height.

$$\frac{\Delta}{h} = \frac{5.1\text{ cm}}{(6.1\text{ m})\left(100\,\frac{cm}{m}\right)}$$
$$= 0.0083$$

*Customary U.S. solution*

Per the UBC [Sec. 1627], *story drift ratio* is the story drift divided by the story height.

$$\frac{\Delta}{h} = \frac{2\text{ in}}{(20\text{ ft})\left(12\,\frac{in}{ft}\right)}$$
$$= 0.0083$$

**62. Answer B**
Structures should be designed to resist the overturning effects caused by seismic loading. When the overturning moment increases faster than the restoring moment from the frame stiffness, the frame is unstable. Per the UBC [Sec. 1630.8.1], overturning effects on every element should be transferred to the foundation.

**63. Answer B**
Based on UBC Table 23-II-H, the maximum allowable shear stress is 360 lbf/ft (5254 N/m).

**64. Answer B**
The allowable shear needs to be equal or greater than the calculated shear of 690 lbf/ft (10 070 N/m). The nail spacing scheme that will give the required shear resistance is 2.5 in (64 mm) spacing at the boundaries. Per UBC Table 23-II-H, this will give a maximum allowable shear stress of 720 lbf/ft (10 508 N/m).

## 65. Answer C

The allowable shear in the shear walls must be equal or greater than the calculated value of 380 lbf/ft (5 546 N/m). Per UBC Table 23-II-I-1, a maximum allowable shear stress of 430 lbf/ft (6 275 N/m) will be achieved when the nail spacing at the wood structural panel edges is 4 in (102 mm).

## 66. Answer D

Based on the UBC [Sec. 1630.8.1], design overturning moment should be distributed to the vertical resisting elements and then carried down to the foundation. The overturning moment is resisted by the weight of the walls (parallel walls only) and by the distributed roof dead load tributary to the walls. Thus, proper actions to resist overturning moment include anchoring the wall panels to the foundation, and increasing the weight of the walls. In addition, larger walls will have a longer resisting moment arm, so making walls longer (or connecting tilt-up panels such that they may be considered as one wall) will increase the resisting moment.

## 67. Answer D

*SI solution*

Based on UBC Table 21-N, for 19 mm diameter bolts in grouted masonry, the allowable shear is 4893 N.

Ignore bolts required for uplift. The number of bolts required is

$$\frac{48\,930 \text{ N}}{4893 \dfrac{\text{N}}{\text{bolt}}} = 10 \text{ bolts}$$

*Customary U.S. solution*

Based on the UBC Table 21-N, for $\frac{3}{4}$ in diameter bolts in grouted masonry, the allowable shear is 1100 lbf.

Ignore bolts required for uplift. The number of bolts required is

$$\frac{11{,}000 \text{ lbf}}{1100 \dfrac{\text{lbf}}{\text{bolt}}} = 10 \text{ bolts}$$

## 68. Answer B

*SI solution*

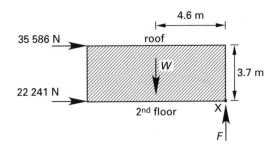

Consider only the moment due to lateral forces applied above location X.

$$\vartheta = \frac{V}{b}$$
$$= \frac{35\,586}{9.1 \text{ m}}$$
$$= 3910.5 \text{ N/m}$$
$$F = \left(3910.5 \; \frac{\text{N}}{\text{m}}\right)(3.7 \text{ m})$$
$$= 14\,468.9$$

The overturning moment is resisted by the dead load of the wall. The 1994 UBC specified that a factor of 0.85 be applied to the dead load in determining the restoring moment. However, in the most recent UBC, no value is explicitly specified. Apparently, it was the intention of the SEAOC "Blue Book" that a factor of 0.9 should be used in load combination. The load combination of the UBC [Sec. 1612.3.1] uses a factor of 0.9. Thus, use a value of 0.9.

$$D_{\text{wall}} = \left(766 \; \frac{\text{N}}{\text{m}^2}\right)(3.7 \text{ m})$$
$$= 2834.2 \text{ N/m}$$

The resisting force is

$$\left(\frac{\left(2834.2 \; \dfrac{\text{N}}{\text{m}}\right)(9.1 \text{ m})}{2}\right)(0.90) = 11\,606 \text{ N}$$

The anchorage force is

$$14\,468.9 \text{ N} - 11\,606 = 2862.9 \text{ N}$$

*Customary U.S. solution*

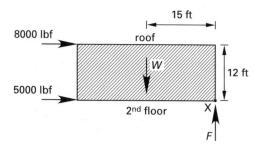

Consider only the moment due to lateral forces applied above location X.

$$\vartheta = \frac{V}{b}$$
$$= \frac{8000 \text{ lbf}}{30 \text{ ft}}$$
$$= 266.7 \text{ lbf/ft}$$

$$F = \left(266.7 \frac{\text{lbf}}{\text{ft}}\right)(12 \text{ ft})$$
$$= 3200 \text{ lbf}$$

The overturning moment is resisted by the dead load of the wall. The 1994 UBC specified that a factor of 0.85 be applied to the dead load in determining the restoring moment. However, in the most recent UBC, no value is explicitly specified. Apparently, it was the intention of the SEAOC "Blue Book" that a factor of 0.9 should be used in load combination. The load combination of the UBC [Sec. 1612.3.1] uses a factor of 0.9. Thus, use a value of 0.9.

$$D_{\text{wall}} = \left(16 \frac{\text{lbf}}{\text{ft}^2}\right)(12 \text{ ft})$$
$$= 192 \text{ lbf/ft}$$

The resisting force is

$$\left(\frac{\left(192 \frac{\text{lbf}}{\text{ft}}\right)(30 \text{ ft})}{2}\right)(90\%) = 2592 \text{ lbf}$$

The anchorage force is

$$3200 \text{ lbf} - 2592 \text{ lbf} = 608 \text{ lbf}$$

69. Answer $\not{D}$

*SI solution*

The required shear capacity is

$$\left(8756 \frac{\text{N}}{\text{m}}\right)(1.2 \text{ m}) = 10\,507 \text{ N per bolt}$$

Based on the UBC [Sec. 2318.2], allowable lateral design values ($Z$) in shear should be as set forth in UBC Tables 23-III-B-1 and 23-III-B-2. These tables give $Z$ values for single shear and double shear for wood-to-wood connections, respectively.

Per the UBC [Sec. 2316.2, Amendment 24 (Sec. 8.3)], allowable shear values for bolts used to connect a wood member to concrete or masonry can be $\frac{1}{2}$ of the double shear values tabulated in UBC Table 23-III-B-2 for a wood member twice the thickness of the member attached to the concrete or masonry. Therefore, the required shear capacity would be

$$V = (10\,507 \text{ N})(2)$$
$$= 21\,014 \text{ N per bolt}$$

Also per the UBC, allowable stresses specified in this code may be increased by $\frac{1}{3}$. Therefore, the required shear capacity would be

$$V = \frac{21\,014 \text{ N}}{1.33}$$
$$= 15\,800 \text{ N per bolt}$$

To determine the length of the bolt, use twice the thickness of the wood member. Therefore,

$$\left(\begin{array}{c}76 \text{ mm nominal} \\ \text{ledger thickness}\end{array}\right)(2) = 152 \text{ mm Douglas fir-larch}$$

Use double shear applied parallel to the grain. Per UBC Table 23-III-B-2, the required shear capacity of 15 800 N per bolt can be achieved with a 25 mm diameter bolt that is 140 mm in length.

The strength of the 25 mm bolt is 18 200 N, which complies with the UBC requirement.

*Customary U.S. solution*

The required shear capacity is

$$\left(600 \frac{\text{lbf}}{\text{ft}}\right)(4 \text{ ft}) = 2400 \text{ lbf per bolt}$$

Based on the UBC [Sec. 2318.2], allowable lateral design values ($Z$) in pounds for bolts in shear should be as set forth in UBC Tables 23-III-B-1 and 23-III-B-2. These tables give $Z$ values for single shear and double shear for wood-to-wood connections, respectively.

Per the UBC [Sec. 2316.2, Amendment 24 (Sec. 8.3)], allowable shear values for bolts used to connect a wood member to concrete or masonry can be $\frac{1}{2}$ of the double shear values tabulated in UBC Table 23-III-B-2 for a wood member twice the thickness of the member attached to the concrete or masonry. Therefore, the required shear capacity would be

$$V = (2400 \text{ lbf})(2)$$
$$= 4800 \text{ lbf per bolt}$$

Also per the UBC, allowable stresses specified in this code may be increased by $\frac{1}{3}$. Therefore, the required shear capacity would be

$$V = \frac{4800 \text{ lbf}}{1.33}$$
$$= 3600 \text{ lbf per bolt}$$

To determine the length of the bolt, use twice the thickness of the wood member. Therefore,

$$\left( \begin{array}{c} 3 \text{ in nominal} \\ \text{ledger thickness} \end{array} \right) (2) = 6 \text{ in Douglas fir-larch}$$

Use double shear applied parallel to the grain. Per UBC Table 23-III-B-2, the required shear capacity of 3600 lbf per bolt can be achieved with a 1 in diameter bolt that is $5\frac{1}{2}$ in in length.

The strength of the 1 in bolt is 4090 lbf, which complies with the UBC requirement.

70. Answer B
Rigidity, $R$, and deflection, $\Delta$, are the reciprocals of each other.

71. Answer D
Rigidity is stiffness; it is a measure of resistance to rotation. For a wall fixed at the bottom and top, the rigidity due to bending and shear is

$$\frac{1}{R_{\text{fixed}}} = \frac{Fh^3}{12EI} + \frac{1.2Fh}{AE_G}$$
$$A = td$$
$$I = \frac{td^3}{12}$$

The rigidity of a wall element depends on its depth, $d$, thickness, $t$, height, $h$, the modulus of elasticity, $E$, the shear modulus, $E_G$, and the conditions of its rotation.

72. Answer B
Increasing the height-to-width ratio of any pier will decrease its rigidity and increase its deflections. Walls can be considered either fixed (i.e., fixed at the top and bottom) or cantilevered (i.e., fixed at the bottom and free at the top). Relative rigidities can be calculated using arbitrary values of shear, wall thickness, and modulus of elasticity. For instance, if $t = 1$ in (25 mm), $F = 100,000$ lbf (444 820 N), and $E = 1 \times 10^6$ lbf/in$^2$ (6.9 $\times$ $10^6$ kPa), the rigidities are given by

$$R_{\text{fixed}} = \frac{1}{(0.1)\left(\dfrac{h}{d}\right)^3 + (0.3)\left(\dfrac{h}{d}\right)}$$

$$R_{\text{cantilever}} = \frac{1}{(0.4)\left(\dfrac{h}{d}\right)^3 + (0.3)\left(\dfrac{h}{d}\right)}$$

$h/d$ is the height-to-width ratio. The larger the $h/d$ ratio, the smaller the relative rigidity. The smaller the relative rigidity, the larger the deflection.

73. Answer B

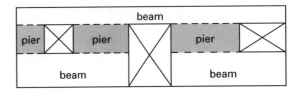

For concrete or masonry walls with openings (e.g., doors or windows), piers refer to the solid vertical portions of the wall. The height is taken as the smaller of the heights of the adjacent openings. Beams are horizontal portions remaining after the piers have been identified. They are above and below the openings. For the wall shown, there are three piers.

74. Answer C
Panels I and II are in parallel. Therefore,

$$\frac{1}{\Delta_{\text{total}}} = \frac{1}{\Delta_{\text{I}}} + \frac{1}{\Delta_{\text{II}}} = \frac{\Delta_{\text{I}} + \Delta_{\text{II}}}{\Delta_{\text{I}}\Delta_{\text{II}}}$$

$$\Delta_{\text{total}} = \frac{\Delta_{\text{I}}\Delta_{\text{II}}}{\Delta_{\text{I}} + \Delta_{\text{II}}}$$

**75. Answer B**

Panels I and II are grouped in series. The deflection of each panel contributes to the deflection of the other panel. Therefore, $\Delta_{\text{total}} = \Delta_{\text{I}} + \Delta_{\text{II}}$.

**76. Answer D**

*SI solution*

The answer can be arrived at intuitively by reasoning that, all other factors being equal, the openings in the north wall will reduce the rigidity and increase the deflections. Calculating the rigidities and deflections will confirm this hypothesis. Use rigidity and deflection tables (you can find these tables in *Seismic Design of Building Structures* by Michael R. Lindeburg, PE).

For the south wall,

$$\frac{h}{d} = \frac{6.1 \text{ m}}{24.4 \text{ m}} = 0.25$$

Therefore, $R = 13.061$ and

$$\Delta_{\text{south wall}} = \frac{25.4}{13.061}$$
$$= 1.9 \text{ mm}$$

For the north wall, panel I,

$$\frac{h_{\text{I}}}{d_{\text{I}}} = \frac{6.1 \text{ m}}{12.2 \text{ m}} = 0.50$$

Therefore, $R_{\text{I}} = 6.154$.

For the north wall, panel II,

$$\frac{h_{\text{II}}}{d_{\text{II}}} = \frac{6.1 \text{ m}}{3.0 \text{ m}} \approx 2.00$$

Therefore, $R_{\text{III}} = 0.714$.

$$R_{\text{total}} = R_{\text{I}} + R_{\text{II}} + R_{\text{III}}$$
$$= 6.154 + 0.714 + 0.714$$
$$= 7.582$$
$$\Delta_{\text{north wall}} = \frac{25.4}{7.582}$$
$$= 3.5 \text{ mm}$$

Thus, $R_{\text{south wall}} > R_{\text{north wall}}$ and $\Delta_{\text{north wall}} > \Delta_{\text{south wall}}$. Note that the openings reduce the rigidity of the walls.

*Customary U.S. solution*

The answer can be arrived at intuitively by reasoning that, all other factors being equal, the openings in the north wall will reduce the rigidity and increase the deflections. Calculating the rigidities and deflections will confirm this hypothesis. Use rigidity and deflection tables (you can find these tables in *Seismic Design of Building Structures* by Michael R. Lindeburg, PE).

For the south wall,

$$\frac{h}{d} = \frac{20 \text{ ft}}{80 \text{ ft}} = 0.25$$

Therefore, $R = 13.061$ and $\Delta_{\text{south wall}} = 0.07 \text{ in}$.

For the north wall, panel I,

$$\frac{h_{\text{I}}}{d_{\text{I}}} = \frac{20 \text{ ft}}{40 \text{ ft}} = 0.50$$

Therefore, $R_{\text{I}} = 6.154$.

For the north wall, panel II,

$$\frac{h_{\text{II}}}{d_{\text{II}}} = \frac{20 \text{ ft}}{10 \text{ ft}} = 2.00$$

Therefore, $R_{\text{II}} = 0.714$.

For the north wall, panel III,

$$\frac{h_{\text{III}}}{d_{\text{III}}} = \frac{20 \text{ ft}}{10 \text{ ft}} = 2.00$$

Therefore, $R_{\text{III}} = 0.714$.

$$R_{\text{total}} = R_{\text{I}} + R_{\text{II}} + R_{\text{III}}$$
$$= 6.154 + 0.714 + 0.714$$
$$= 7.582$$
$$\Delta_{\text{north wall}} = 0.13 \text{ in}$$

Thus, $R_{\text{south wall}} > R_{\text{north wall}}$ and $\Delta_{\text{north wall}} > \Delta_{\text{south wall}}$. Note that the openings reduce the rigidity of the walls.

**77. Answer C**

*SI solution*

Use rigidity and deflection tables.

For wall I,

$$\frac{h}{d} = \frac{L}{2L} = 0.50$$

Therefore, $R = 6.15$ and

$$\Delta_{\mathrm{I}} = \frac{25.4}{6.15}$$
$$= 4.1 \text{ mm}$$

For wall II,
$$\frac{h}{d} = \frac{L}{2L} = 0.50$$

Therefore, $R = 6.15$ and

$$\Delta_{\mathrm{II}} = \frac{25.4}{6.15}$$
$$= 4.1 \text{ mm}$$

For wall III, panel 1,

$$\frac{h}{d} = \frac{L}{L} = 1.00$$

Therefore, $R = 2.50$ and

$$\Delta_{\mathrm{I}} = \frac{25.4}{2.5}$$
$$= 10.2 \text{ mm}$$

For wall III, panel 2, $h/d = L/L = 1.00$. Therefore, $R = 2.50$ and $\Delta_2 = 10.2$ mm.

Panels 1 and 2 are in parallel. Therefore,

$$\frac{1}{\Delta_{\mathrm{III}}} = \frac{1}{\Delta_1} + \frac{1}{\Delta_2}$$
$$= \frac{1}{10.2} + \frac{1}{10.2}$$
$$\Delta_{\mathrm{III}} = 5.10 \text{ mm}$$

Thus, $\Delta_{\mathrm{III}}$ is greater than $\Delta_{\mathrm{I}}$ and $\Delta_{\mathrm{II}}$. Wall III has the highest deflection.

*Customary U.S. solution*

Use rigidity and deflection tables.

For wall I,
$$\frac{h}{d} = \frac{L}{2L} = 0.50$$

Therefore, $R = 6.15$ and $\Delta_{\mathrm{I}} = 0.16$ in.

For wall II,
$$\frac{h}{d} = \frac{L}{2L} = 0.50$$

Therefore, $R = 6.15$, and $\Delta_{\mathrm{II}} = 0.16$ in.

For wall III, panel 1,

$$\frac{h}{d} = \frac{L}{L} = 1.00$$

Therefore, $R = 2.50$ and $\Delta_1 = 0.4$ in.

For wall III, panel 2,

$$\frac{h}{d} = \frac{L}{L} = 1.00$$

Therefore, $R = 2.50$, and $\Delta_2 = 0.4$ in.

Panels 1 and 2 are in parallel. Therefore,

$$\frac{1}{\Delta_{\mathrm{III}}} = \frac{1}{\Delta_1} + \frac{1}{\Delta_2}$$
$$= \frac{1}{0.4} + \frac{1}{0.4}$$
$$\Delta_{\mathrm{III}} = 0.2 \text{ in}$$

Thus, $\Delta_{\mathrm{III}}$ is greater than $\Delta_{\mathrm{I}}$ and $\Delta_{\mathrm{II}}$. Wall III has the highest deflection.

**78. Answer B**

*SI solution*

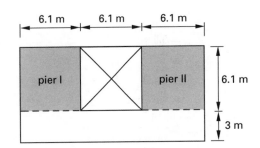

The method used to approximate the rigidity of the wall involves calculating the deflection of the wall as if it were solid, subtracting the deflection of a wall that has a height equal to the height of the opening, and then adding in the deflections of the piers within the removed strip. Use rigidity and deflection tables for fixed piers.

For the solid wall (assuming no opening),

$$\frac{h}{d} = \frac{9.1 \text{ m}}{18.3 \text{ m}} = 0.50$$
$$R = 6.15$$
$$\Delta = \frac{25.4}{R}$$
$$= 4.1 \text{ mm}$$

For the strip wall with an opening,

$$\frac{h}{d} = \frac{6.1 \text{ m}}{18.3 \text{ m}} = 0.33$$
$$R = 9.747$$
$$\Delta = \frac{25.4}{R} = \frac{25.4}{9.747}$$
$$= 2.6 \text{ mm}$$

For pier I,

$$\frac{h}{d} = \frac{6.1 \text{ m}}{6.1 \text{ m}} = 1.00$$
$$R = 2.50$$

For pier II,

$$\frac{h}{d} = \frac{6.1 \text{ m}}{6.1 \text{ m}} = 1.00$$
$$R = 2.50$$

The piers are in parallel. Therefore,

$$R_{\text{total}} = R_{\text{I}} + R_{\text{II}}$$
$$= 2.5 + 2.5$$
$$= 5.0$$
$$\Delta = \frac{25.4}{R} = \frac{25.4}{5.0}$$
$$= 5.1 \text{ mm}$$
$$\Delta_{\text{wall}} = \Sigma\Delta$$
$$= 4.1 + (-2.6) + 5.1$$
$$= 6.6 \text{ mm}$$
$$R_{\text{wall}} = \frac{1}{\Delta_{\text{wall}}}$$
$$R_{\text{wall}} = \frac{25.4}{6.6 \text{ mm}} = 3.84 \approx 4.0$$

*Customary U.S. solution*

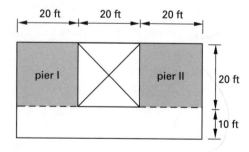

The method used to approximate the rigidity of the wall involves calculating the deflection of the wall as if it were solid, subtracting the deflection of a wall that has a height equal to the height of the opening, and then adding in the deflections of the piers within the removed strip. Use rigidity and deflection tables for fixed piers.

For the solid wall (assuming no opening),

$$\frac{h}{d} = \frac{30 \text{ ft}}{60 \text{ ft}} = 0.50$$
$$R = 6.15$$
$$\Delta = 0.16 \text{ in}$$

For the strip wall with an opening,

$$\frac{h}{d} = \frac{20 \text{ ft}}{60 \text{ ft}} = 0.33$$
$$R = 9.747$$
$$\Delta = 0.10 \text{ in}$$

For pier I,

$$\frac{h}{d} = \frac{20 \text{ ft}}{20 \text{ ft}} = 1.00$$
$$R = 2.50$$

For pier II,

$$\frac{h}{d} = \frac{20 \text{ ft}}{20 \text{ ft}} = 1.00$$
$$R = 2.50$$

The piers are in parallel. Therefore,

$$R_{\text{total}} = R_{\text{I}} + R_{\text{II}}$$
$$= 2.5 + 2.5$$
$$= 5.0$$
$$\Delta = \frac{1}{R} = \frac{1}{5.0}$$
$$= 0.2 \text{ in}$$
$$\Delta_{\text{wall}} = \Sigma\Delta$$
$$= 0.16 + (-0.10) + 0.2$$
$$= 0.26 \text{ in}$$
$$R_{\text{wall}} = \frac{1}{\Delta_{\text{wall}}}$$
$$R_{\text{wall}} = \frac{1}{0.26 \text{ in}} = 3.84$$
$$= 4$$

**79. Answer D**

*SI solution*

The rigidity is the reciprocal of the deflection.

$$R = \frac{25.4}{\Delta \text{ mm}}$$
$$\Delta = 2.5 \text{ mm}$$
$$R = \frac{25.4}{2.5 \text{ mm}} = 10$$

*Customary U.S. solution*

The rigidity is the reciprocal of the deflection.

$$R = \frac{1}{\Delta \text{ in}} = \frac{1}{0.1 \text{ in}} = 10$$

**80. Answer C**

*SI solution*

The rigidity is the reciprocal of the deflection. The rigidity of the entire wall is equal to the sum of the component rigidities.

$$R_{\text{steel}} = \frac{25.4}{\Delta_{\text{steel frame}}}$$
$$\Delta = 25 \text{ mm}$$
$$R_{\text{steel}} = \frac{25.4}{25 \text{ mm}}$$
$$= 1.0$$
$$R_{\text{wall}} = R_{\text{shear wall}} + R_{\text{steel frame}}$$
$$R_{\text{shear wall}} = 10.0 - 1.0$$
$$= 9.0$$

The rigidity of the steel frame is 1.0, and the rigidity of the masonry shear wall is 9.0.

*Customary U.S. solution*

The rigidity is the reciprocal of the deflection. The rigidity of the entire wall is equal to the sum of the component rigidities.

$$R_{\text{steel}} = \frac{1}{\Delta_{\text{steel frame}}} = \frac{1}{1 \text{ in}}$$
$$= 1.0$$
$$R_{\text{wall}} = R_{\text{shear wall}} + R_{\text{steel frame}}$$
$$R_{\text{shear wall}} = 10.0 - 1.0$$
$$= 9.0$$

The rigidity of the steel frame is 1.0, and the rigidity of the masonry shear wall is 9.0.

**81. Answer A**

*SI solution*

The steel frame and shear wall resist the lateral load in proportion to their rigidities.

$$R_{\text{wall}} = R_{\text{shear wall}} + R_{\text{shear frame}}$$
$$= 9R_{\text{shear wall}} + R_{\text{shear wall}}$$
$$= 10R_{\text{shear wall}}$$
$$V_{\text{steel frame}} = \left( \frac{R_{\text{steel frame}}}{R_{\text{steel frame}} + \left( \frac{1}{9} \right) \left( R_{\text{steel frame}} \right)} \right) (267 \text{ kN})$$
$$= (0.9)(267 \text{ kN})$$
$$= 240 \text{ kN}$$
$$V_{\text{shear wall}} = \left( \frac{R_{\text{shear wall}}}{R_{\text{wall}}} \right) (267 \text{ kN})$$
$$= \left( \frac{R_{\text{shear wall}}}{10R_{\text{shear wall}}} \right) (267 \text{ kN})$$
$$= (0.1)(267 \text{ kN})$$
$$= 27 \text{ kN}$$

*Customary U.S. solution*

The steel frame and shear wall resist the lateral load in proportion to their rigidities.

$$R_{\text{wall}} = R_{\text{shear wall}} + R_{\text{steel frame}}$$
$$= 9R_{\text{shear wall}} + R_{\text{shear wall}}$$
$$= 10R_{\text{shear wall}}$$
$$V_{\text{steel frame}} = \left( \frac{R_{\text{steel frame}}}{R_{\text{steel frame}} + \left( \frac{1}{9} \right) \left( R_{\text{steel frame}} \right)} \right) (60 \text{ k})$$
$$= (0.9)(60 \text{ k})$$
$$= 54 \text{ k}$$
$$V_{\text{shear wall}} = \left( \frac{R_{\text{shear wall}}}{R_{\text{wall}}} \right) (60 \text{ k})$$
$$= \left( \frac{R_{\text{shear wall}}}{10R_{\text{shear wall}}} \right) (60 \text{ k})$$
$$= (0.1)(60 \text{ k})$$
$$= 6 \text{ k}$$

82. Answer D

Rigid diaphragms distribute loads to resisting vertical elements in proportion to their rigidities. Since the rigidity of each panel is the same, each panel will carry the same load.

$$R_{\mathrm{I}} = R_{\mathrm{II}}$$
$$R_{\mathrm{total}} = R_{\mathrm{I}} + R_{\mathrm{II}} = 2R_{\mathrm{I}}$$
$$= 2R_{\mathrm{II}}$$
$$V_{\mathrm{I}} = \left(\frac{R_{\mathrm{I}}}{2R_{\mathrm{I}}}\right)F = \frac{F}{2}$$
$$V_{\mathrm{II}} = \left(\frac{R_{\mathrm{II}}}{2R_{\mathrm{II}}}\right)F = \frac{F}{2}$$
$$V_{\mathrm{I}} = V_{\mathrm{II}}$$

Therefore, both panels carry an equal load.

83. Answer C

*SI solution*

The load will be distributed to the vertical elements in proportion to their rigidities.

For the steel frame,

$$R = \frac{25.4}{\Delta \text{ mm}}$$
$$\Delta = 25 \text{ mm}$$
$$R = \frac{25.4}{25 \text{ mm}} = 1$$

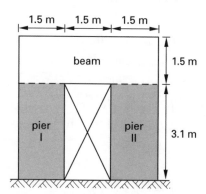

To approximate the rigidity of the masonry shear wall, calculate the deflection. First, assume the wall to be solid, then subtract the deflection of a wall with height equal to the opening. Then add the deflection contributed by the piers in the removed section.

For a solid wall,

$$\frac{h}{d} = \frac{4.5 \text{ m}}{4.5 \text{ m}} = 1.00$$
$$R = 2.50$$
$$\Delta = \frac{25.4}{R}$$
$$= \frac{25.4}{2.5}$$
$$= 10.2 \text{ mm}$$

For a strip wall with opening,

$$\frac{h}{d} = \frac{3.1 \text{ m}}{4.5 \text{ m}} = 0.68$$
$$R = 4.25$$
$$\Delta = \frac{1}{R} = \frac{1}{4.25} = \frac{25.4}{R}$$
$$= \frac{25.4}{4.25}$$
$$= 6.0 \text{ mm}$$

For pier I,

$$\frac{h}{d} = \frac{3.1 \text{ m}}{1.5 \text{ m}} = 2.00$$
$$R = 0.71$$

For pier II,

$$\frac{h}{d} = \frac{3.1 \text{ m}}{1.5 \text{ m}} = 2.00$$
$$R = 0.71$$

The piers are in parallel. Therefore,

$$R_{\mathrm{total}} = R_{\mathrm{I}} + R_{\mathrm{II}} = 0.71 + 0.71$$
$$= 1.42$$
$$\Delta = \frac{25.4}{R}$$
$$= \frac{25.4}{1.42}$$
$$= 17.8 \text{ mm}$$
$$\Delta_{\mathrm{wall}} = \Sigma\Delta = 10.2 - 6.0 + 17.8$$
$$= 22.0 \text{ mm}$$
$$R_{\mathrm{wall}} = \frac{25.4}{\Delta_{\mathrm{wall}}}$$
$$= \frac{25.4}{22.0 \text{ mm}}$$
$$= 1.15$$
$$R_{\mathrm{west\ wall}} = R_{\mathrm{steel\ frame}} + R_{\mathrm{shear\ wall}} + R_{\mathrm{steel\ frame}}$$
$$= 1 + 1.15 + 1$$
$$= 3.15$$

The lateral load carried by the masonry shear wall is

$$V_{\text{shear wall}} = \left(\frac{R_{\text{shear wall}}}{R_{\text{west wall}}}\right) V_{\text{total}}$$

$$= \left(\frac{1.15}{3.15}\right)(267 \text{ kN})$$

$$= 97.5 \text{ kN}$$

*Customary U.S. solution*

The load will be distributed to the vertical elements in proportion to their rigidities.

For the steel frame,

$$R = \frac{1}{\Delta} = \frac{1}{1 \text{ in}} = 1$$

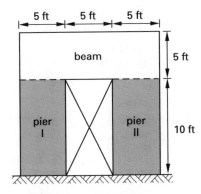

To approximate the rigidity of the masonry shear wall, calculate the deflection. First, assume the wall to be solid, then subtract the deflection of a wall with height equal to the opening. Then add the deflection contributed by the piers in the removed section.

For a solid wall,

$$\frac{h}{d} = \frac{15 \text{ ft}}{15 \text{ ft}} = 1.00$$

$$R = 2.50$$

$$\Delta = 0.40 \text{ in}$$

For a strip wall with opening,

$$\frac{h}{d} = \frac{10 \text{ ft}}{15 \text{ ft}} = 0.67$$

$$R = 4.33$$

$$\Delta = 0.23 \text{ in}$$

For pier I,

$$\frac{h}{d} = \frac{10 \text{ ft}}{5 \text{ ft}} = 2.00$$

$$R = 0.71$$

For pier II,

$$\frac{h}{d} = \frac{10 \text{ ft}}{5 \text{ ft}} = 2.00$$

$$R = 0.71$$

The piers are in parallel. Therefore,

$$R_{\text{total}} = R_{\text{I}} + R_{\text{II}} = 0.71 + 0.71$$

$$= 1.42$$

$$\Delta = \frac{1}{R} = \frac{1}{1.42}$$

$$= 0.70 \text{ in}$$

$$\Delta_{\text{wall}} = \Sigma\Delta = 0.40 + (-0.23) + 0.70$$

$$= 0.87 \text{ in}$$

$$R_{\text{wall}} = \frac{1}{\Delta_{\text{wall}}}$$

$$R_{\text{wall}} = \frac{1}{0.87}$$

$$= 1.15$$

$$R_{\text{west wall}} = R_{\text{steel frame}} + R_{\text{shear wall}} + R_{\text{steel frame}}$$

$$= 1 + 1.15 + 1$$

$$= 3.15$$

The lateral load carried by the masonry shear wall is

$$V_{\text{shear wall}} = \left(\frac{R_{\text{shear wall}}}{R_{\text{west wall}}}\right) V_{\text{total}}$$

$$= \left(\frac{1.15}{3.15}\right)(60 \text{ k})$$

$$= 21.9 \text{ k}$$

## 84. Answer D

*SI solution*

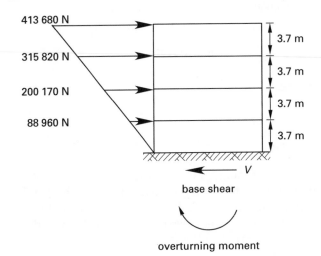

The overturning moment is the sum of the moments about the base due to the distributed shear at each floor.

$$(88\,960 \text{ N})(3.7 \text{ m}) + (200\,170 \text{ N})(7.4 \text{ m})$$
$$+ (315\,820 \text{ N})(11.1 \text{ m}) + (413\,680 \text{ N})(14.8 \text{ m})$$
$$= 11\,438\,476 \text{ N·m}$$
$$\approx 11\,440 \text{ kN·m}$$

*Customary U.S. solution*

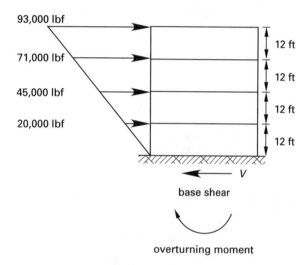

base shear

overturning moment

The overturning moment is the sum of the moments about the base due to the distributed shear at each floor.

$$(20{,}000 \text{ lbf})(12 \text{ ft}) + (45{,}000 \text{ lbf})(24 \text{ ft})$$
$$+ (71{,}000 \text{ lbf})(36 \text{ ft}) + (93{,}000 \text{ lbf})(48 \text{ ft})$$
$$= 8{,}340{,}000 \text{ ft-lbf}$$
$$= 8340 \text{ ft-k}$$

**85. Answer B**

*SI solution*

The walls should be designed to resist the sum of the diaphragm force and the inertia force of the parallel walls. Only the top half of the walls contribute their weight to the inertia force; the bottom half is considered to be transferred to the foundation.

The weight of the shear wall is

$$W_{\text{shear wall}} = \left(\frac{4.9 \text{ m}}{2}\right)(15.2 \text{ m})\left(766 \ \frac{\text{N}}{\text{m}^2}\right)$$
$$= 28\,526 \text{ N}$$
$$V_{\text{shear wall}} = \left(\frac{C_v I}{RT}\right)W$$
$$= (0.1375)(28\,526 \text{ N})$$
$$= 3922 \text{ N}$$
$$V_{\text{total shear}} = V_{\text{shear wall}} + V_{\text{roof}}$$
$$= 44\,482 \text{ N} + 3922 \text{ N}$$
$$= 48\,404 \text{ N}$$
$$\vartheta = \frac{V}{b}$$
$$= \frac{48\,404 \text{ N}}{15.2 \text{ m}}$$
$$= 3184.5 \text{ N/m}$$

*Customary U.S. solution*

The walls should be designed to resist the sum of the diaphragm force and the inertia force of the parallel walls. Only the top half of the walls contribute their weight to the inertia force; the bottom half is considered to be transferred to the foundation.

The weight of the shear wall is

$$W_{\text{shear wall}} = \left(\frac{16 \text{ ft}}{2}\right)(50 \text{ ft})\left(16 \ \frac{\text{lbf}}{\text{ft}^2}\right)$$
$$= 6400 \text{ lbf}$$
$$V_{\text{shear wall}} = \left(\frac{C_v I}{RT}\right)W$$
$$= (0.1375)(6400 \text{ lbf})$$
$$= 880 \text{ lbf}$$
$$V_{\text{total shear}} = V_{\text{shear wall}} + V_{\text{roof}}$$
$$= 10{,}000 + 880$$
$$= 10{,}880 \text{ lbf}$$
$$\vartheta = \frac{V}{b}$$
$$= \frac{10{,}880 \text{ lbf}}{50 \text{ ft}}$$
$$= 217.6 \text{ lbf/ft}$$

**86.** Answer D

*SI solution*

Resistance against overturning is provided by the dead loads of the walls and roof. In this case, the contribution from the roof is neglected.

$$D_{\text{wall}} = (12.2 \text{ m})(4.6 \text{ m}) \left( 958 \, \frac{\text{N}}{\text{m}^2} \right)$$
$$= 53\,763.0 \text{ N}$$

Per the UBC [Sec. 1633.1], consideration should be given to design for uplift effects caused by seismic loads. To reduce uplift, the 1994 UBC limited the dead load contribution to 85%. However, in the most recent UBC, no value is explicitly specified. Apparently, it was the intention of the SEAOC "Blue Book" that a factor of 0.9 be used in load combination. The load combination of the UBC [Sec. 1612.3.1] uses a factor of 0.9. Thus, use a value of 0.9. The pivot point is at the corner and base of the wall, so the moment arm for the resisting force is $L/2$.

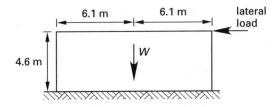

The resisting moment is

$$\frac{(0.9)(53\,763 \text{ N})(12.2 \text{ m})}{2} = 295\,158.8 \text{ N·m}$$

*Customary U.S. solution*

Resistance against overturning is provided by the dead loads of the walls and roof. In this case, the contribution from the roof is neglected.

$$D_{\text{wall}} = (40 \text{ ft})(15 \text{ ft}) \left( 20 \, \frac{\text{lbf}}{\text{ft}^2} \right)$$
$$= 12{,}000 \text{ lbf}$$

Per the UBC [Sec. 1633.1], consideration should be given to design for uplift effects caused by seismic loads. To reduce uplift, the 1994 UBC limited the dead load contribution to 85%. However, in the most recent UBC,

no value is explicitly specified. Apparently, it was the intention of the SEAOC "Blue Book" that a factor of 0.9 be used in load combination. The load combination of the UBC [Sec. 1612.3.1] uses a factor of 0.9. Thus, use a value of 0.9. The pivot point is at the corner and base of the wall, so the moment arm for the resisting force is $L/2$.

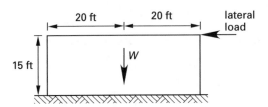

The resisting moment is

$$\frac{(0.9)(12{,}000 \text{ lbf})(40 \text{ ft})}{2} = 216{,}000 \text{ ft-lbf}$$

**87.** Answer D

*SI solution*

The tie-down force is the difference between the calculated overturning moment and the resisting moment.

Calculate the shear due to the diaphragm load. The diaphragm load consists of the inertia effects of the seismic load on the roof and perpendicular walls.

From UBC Table 16-I, $Z = 0.4$. From UBC Table 16-K, $I = 1.0$. From UBC Table 16-N, $R = 5.5$. Based on the UBC [Secs. 1629.3 and 1636.2], when the soil properties are not known in sufficient detail to determine the soil profile type, $S_D$ should be assigned. From UBC Table 16-Q, for the soil profile type $S_D$ and seismic zone 4, the seismic response coefficient $C_a$ is

$$C_a = 0.44 N_a$$
$$= (0.44)(1.0)$$
$$= 0.44$$

For N-S loading,

$$W_{\text{roof}} = (30.5 \text{ m})(3.7 \text{ m})(3) \left( 958 \, \frac{\text{N}}{\text{m}^2} \right)$$
$$= 324\,330.9 \text{ N}$$

$$W_{\text{walls}} = (2 \text{ walls})(30.5 \text{ m}) \left( \frac{4.3 \text{ m}}{2} \right) \left( 766 \, \frac{\text{N}}{\text{m}^2} \right)$$
$$= 100\,460.9 \text{ N}$$

$$W_{\text{total}} = 324\,330.9 \text{ N} + 100\,460.9 \text{ N}$$
$$= 424\,791.8 \text{ N}$$

The base shear should be determined from the simplified design base shear procedure. Use UBC Formula 30-11.

$$V = \left( \frac{3.0 C_a}{R} \right) W$$

$$= \left( \frac{(3.0)(0.44)}{5.5} \right) (424\,791.8\text{ N})$$

$$= 101\,950\text{ N}$$

The shear load carried by each parallel wall is

$$R_{\text{roof}} = \frac{V}{2}$$

$$= \frac{101\,950\text{ N}}{2}$$

$$= 50\,975\text{ N}$$

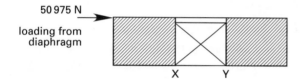

50 975 N

loading from
diaphragm

X          Y

$$\vartheta_R = \frac{V}{b}$$

$$= \frac{50\,975\text{ N}}{7.4\text{ m}}$$

$$= 6888.5\text{ N/m}$$

Next calculate the shear due to the seismic force on the parallel wall. Note that only the top halves of the walls contribute their weight to the inertia force; the weight of the bottom halves is transferred to the foundation.

$$W_{\text{shear wall}} = \left( \frac{4.3\text{ m}}{2} \right) (7.4\text{ m}) \left( 766\ \frac{\text{N}}{\text{m}^2} \right)$$

$$= 12\,187.1\text{ N}$$

$$V = \left( \frac{(3.0)(0.44)}{5.5} \right) (12\,187.1\text{ N})$$

$$= 2924.9\text{ N}$$

$$\vartheta_{\text{wall}} = \frac{V}{b}$$

$$= \frac{2924.9\text{ N}}{7.4\text{ m}}$$

$$= 395.3\text{ N/m}$$

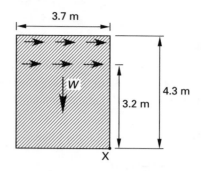

3.7 m

W

4.3 m

3.2 m

X

For the load from the top half portion of the wall,

$$D_{\text{wall}} = \left( 766\ \frac{\text{N}}{\text{m}^2} \right) (4.3\text{ m})$$

$$= 3293.8\text{ N/m}$$

Consideration should be given to design for uplift effects caused by seismic loads. To reduce uplift, the 1994 UBC limited the dead load contribution to 85%. However, no value is explicitly specified in the most recent UBC. Apparently, it was the intention of the SEAOC "Blue Book" that a factor of 0.9 be used in load combination. The load combination of the UBC [Sec. 1612.3.1] uses a factor of 0.9. Thus, use a value of 0.9.

The design tie-down force at X is

$$\left( 6888.5\ \frac{\text{N}}{\text{m}} \right) (4.3\text{ m}) + \left( 395.3\ \frac{\text{N}}{\text{m}} \right) (3.2\text{ m})$$

$$- (0.9) \left( 3293.8\ \frac{\text{N}}{\text{m}} \right) \left( \frac{3.7\text{ m}}{2} \right) = 27\,229.4\text{ N}$$

The tie-down force at Y is identical to the tie-down force at X.

*Customary U.S. solution*

The tie-down force is the difference between the calculated overturning moment and the resisting moment.

Calculate the shear due to the diaphragm load. The diaphragm load consists of the inertia effects of the seismic load on the roof and perpendicular walls.

From UBC Table 16-I, $Z = 0.4$. From UBC Table 16-K, $I = 1.0$. From UBC Table 16-N, $R = 5.5$. Based on the UBC [Secs. 1629.3 and 1636.2], when the soil properties are not known in sufficient detail to determine the soil profile type, $S_D$ should be assigned. From UBC Table 16-Q, for the soil profile type $S_D$ and seismic zone 4, the seismic response coefficient $C_a$ is

$$C_a = 0.44N_a$$
$$= (0.44)(1.0)$$
$$= 0.44$$

For N-S loading,

$$W_{\text{roof}} = (100 \text{ ft})(36 \text{ ft})\left(20 \frac{\text{lbf}}{\text{ft}^2}\right)$$
$$= 72{,}000 \text{ lbf}$$
$$W_{\text{wall}} = (2 \text{ walls})(100 \text{ ft})\left(\frac{14 \text{ ft}}{2}\right)\left(16 \frac{\text{lbf}}{\text{ft}^2}\right)$$
$$= 22{,}400 \text{ lbf}$$
$$W_{\text{total}} = 72{,}000 \text{ lbf} + 22{,}400 \text{ lbf}$$
$$= 94{,}400 \text{ lbf}$$

The base shear should be determined from the simplified design base shear procedure. Use UBC Formula 30-11.

$$V = \left(\frac{3.0C_a}{R}\right)W$$
$$= \left(\frac{(3.0)(0.44)}{5.5}\right)(94{,}400 \text{ lbf})$$
$$= 22{,}656 \text{ lbf}$$

The shear load carried by each parallel wall is

$$R_{\text{roof}} = \frac{V}{2}$$
$$= \frac{22{,}656 \text{ lbf}}{2}$$
$$= 11{,}328 \text{ lbf}$$

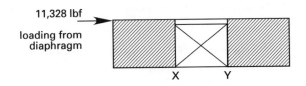

11,328 lbf

loading from diaphragm

X    Y

$$\vartheta_R = \frac{V}{b}$$
$$= \frac{11{,}328 \text{ lbf}}{24 \text{ ft}}$$
$$= 472 \text{ lbf/ft}$$

Next calculate the shear due to the seismic force on the parallel wall. Note that only the top halves of the walls contribute their weight to the inertia force; the weight of the bottom halves is transferred to the foundation.

$$W_{\text{shear wall}} = \left(\frac{14 \text{ ft}}{2}\right)(24 \text{ ft})\left(16 \frac{\text{lbf}}{\text{ft}^2}\right)$$
$$= 2688 \text{ lbf}$$
$$V = \left(\frac{(3.0)(0.44)}{5.5}\right)(2688 \text{ lbf})$$
$$= 645.12 \text{ lbf}$$
$$\vartheta_{\text{wall}} = \frac{V}{b}$$
$$= \frac{645.12 \text{ lbf}}{24 \text{ ft}}$$
$$= 26.9 \text{ lbf/ft}$$

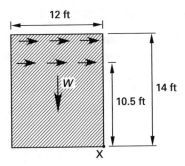

12 ft

W

10.5 ft

14 ft

X

For the load from the top half portion of the wall,

$$D_{\text{wall}} = \left(16 \frac{\text{lbf}}{\text{ft}^2}\right)(14 \text{ ft})$$
$$= 224 \text{ lbf/ft}$$

Consideration should be given to design for uplift effects caused by seismic loads. To reduce uplift, the 1994 UBC limited the dead load contribution to 85%. However, no value is explicitly specified in the most recent UBC. Apparently, it was the intention of the SEAOC "Blue Book" that a factor of 0.9 be used in load combination. The load combination of the UBC [Sec. 1612.3.1] uses a factor of 0.9. Thus, use a value of 0.9.

The design tie-down force at X is

$$\left(472 \frac{\text{lbf}}{\text{ft}}\right)(14 \text{ ft}) + \left(26.9 \frac{\text{lbf}}{\text{ft}}\right)(10.5 \text{ ft})$$
$$- (0.9)\left(224 \frac{\text{lbf}}{\text{ft}}\right)\left(\frac{12 \text{ ft}}{2}\right) = 5681 \text{ lbf}$$

The tie-down force at Y is identical to the tie-down force at X.

## 88. Answer B

*SI solution*

Buildings with rigid diaphragms distribute the lateral load to the resisting elements in proportion to the rigidities of the elements. The rigidity of the perpendicular wall is neglected (omitting the weak walls).

$$V_{6.1 \text{ m wall}} = \left( \frac{R_{6.1 \text{ m wall}}}{R_{12.2 \text{ m wall}} + R_{6.1 \text{ m wall}}} \right) V_{\text{N-S}}$$
$$= \left( \frac{4}{6+4} \right) \left( 7297 \frac{\text{N}}{\text{m}} \right) (30.5 \text{ m})$$
$$= 89\,023.4 \text{ N}$$
$$\approx 89\,000 \text{ N}$$

*Customary U.S. solution*

Buildings with rigid diaphragms distribute the lateral load to the resisting elements in proportion to the rigidities of the elements. The rigidity of the perpendicular wall is neglected (omitting the weak walls).

$$V_{20 \text{ ft wall}} = \left( \frac{R_{20 \text{ ft wall}}}{R_{40 \text{ ft wall}} + R_{20 \text{ ft wall}}} \right) V_{\text{N-S}}$$
$$= \left( \frac{4}{6+4} \right) \left( 500 \frac{\text{lbf}}{\text{ft}} \right) (100 \text{ ft})$$
$$= 20{,}000 \text{ lbf}$$

## 89. Answer D

*SI solution*

The lateral force is distributed to the walls in proportion to their rigidities.

$$V_{\text{I}} = \left( \frac{R_{\text{I}}}{R_{\text{I}} + R_{\text{II}} + R_{\text{III}}} \right) V_{\text{N-S}}$$
$$= \left( \frac{6}{6+3+5} \right) \left( 7297 \frac{\text{N}}{\text{m}} \right) (30.5 \text{ m})$$
$$= 95\,382.2 \text{ N}$$
$$\approx 95\,400 \text{ N}$$

*Customary U.S. solution*

The lateral force is distributed to the walls in proportion to their rigidities.

$$V_{\text{I}} = \left( \frac{R_{\text{I}}}{R_{\text{I}} + R_{\text{II}} + R_{\text{III}}} \right) V_{\text{N-S}}$$
$$= \left( \frac{6}{6+3+5} \right) (500 \text{ lbf})(100 \text{ ft})$$
$$= 21{,}429 \text{ lbf}$$

## 90. Answer A

Wall I has the highest relative rigidity among all the walls. Therefore, it will resist more of the lateral force than walls II or III.

## 91. Answer C

*SI solution*

This problem can be solved by inspection; $R$ should be equal to 5.

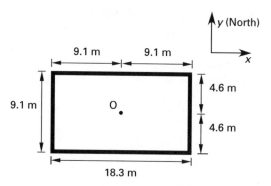

Point O is the center of the diaphragm.

| wall | $R$ | $y$ |
|---|---|---|
| south | ? | −4.6 m |
| north | 5 | +4.6 m |

$$\bar{y}_R = \frac{R_S(-4.6 \text{ m}) + (5)(4.6 \text{ m})}{R_S + 5}$$

With the origin, O, at the center of the diaphragm, $\bar{y}_R$ is zero.

$$(0)(R_S + 5) = R_S(-4.6 \text{ m}) + (5)(+4.6 \text{ m})$$
$$R_S = \frac{(5)(4.6 \text{ m})}{4.6 \text{ m}}$$
$$= 5$$

*Customary U.S. solution*

This problem can be solved by inspection; $R$ should be equal to 5.

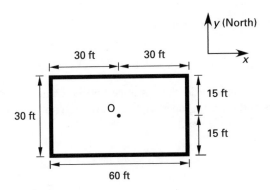

Point O is the center of the diaphragm.

| wall | $R$ | $y$ |
|---|---|---|
| south | ? | −15 ft |
| north | 5 | +15 ft |

$$\overline{y}_R = \frac{R_S(-15 \text{ ft}) + (5)(15 \text{ ft})}{R_S + 5}$$

With the origin, O, at the center of the diaphragm, $y_R$ is zero.

$$(0)(R_S + 5) = R_S(-15 \text{ ft}) + (5)(+15 \text{ ft})$$

$$R_S = \frac{(5)(15 \text{ ft})}{15 \text{ ft}} = 5$$

**92.  Answer C**

*SI solution*

| wall | $R$ | $x$ | $y$ |
|---|---|---|---|
| A | 5 | 7.6 m | 0 m |
| B | 4 | 0 m | 6.1 m |
| C | 3 | 15.2 m | 12.2 m |
| D | 2 | 30.5 m | 6.1 m |
| E | 1 | 29.0 m | 0 m |

For N-S earthquakes, walls B and D are active.

$$\overline{x}_R = \frac{(4)(0 \text{ m}) + (2)(30.4 \text{ m})}{4 + 2}$$

$$= 10.2 \text{ m}$$

$$\overline{y}_R = \frac{(5)(0 \text{ m}) + (1)(0 \text{ m}) + (3)(12.2 \text{ m})}{5 + 1 + 3}$$

$$= 4.1 \text{ m}$$

*Customary U.S. solution*

| wall | $R$ | $x$ | $y$ |
|---|---|---|---|
| A | 5 | 25 ft | 0 ft |
| B | 4 | 0 ft | 20 ft |
| C | 3 | 50 ft | 40 ft |
| D | 2 | 100 ft | 20 ft |
| E | 1 | 95 ft | 0 ft |

For N-S earthquakes, walls B and D are active.

$$\overline{x}_R = \frac{(4)(0 \text{ ft}) + (2)(100 \text{ ft})}{4 + 2}$$

$$= 33.3 \text{ ft}$$

$$\overline{y}_R = \frac{(5)(0 \text{ ft}) + (1)(0 \text{ ft}) + (3)(40 \text{ ft})}{5 + 1 + 3}$$

$$= 13.3 \text{ ft}$$

**93.  Answer C**

*SI solution*

| wall | $R$ | $x$ | $y$ |
|---|---|---|---|
| I | 4 | 4.6 m | 4.6 m |
| II | 2 | 4.6 m | 22.9 m |
| III | 3 | 22.9 m | 22.9 m |
| IV | 3 | 22.9 m | 4.6 m |

$$\overline{x}_R = \frac{(4)(4.6 \text{ m}) + (2)(4.6 \text{ m}) + (3)(22.9 \text{ m}) + (3)(22.9 \text{ m})}{4 + 2 + 3 + 3}$$

$$= 13.7 \text{ m}$$

$$\overline{y}_R = \frac{(4)(4.6 \text{ m}) + (2)(22.9 \text{ m}) + (3)(22.9 \text{ m}) + (3)(4.6 \text{ m})}{4 + 2 + 3 + 3}$$

$$= 12.2 \text{ m}$$

*Customary U.S. solution*

| wall | $R$ | $x$ | $y$ |
|---|---|---|---|
| I | 4 | 15 ft | 15 ft |
| II | 2 | 15 ft | 75 ft |
| III | 3 | 75 ft | 75 ft |
| IV | 3 | 75 ft | 15 ft |

$$\bar{x}_R = \frac{(4)(15 \text{ ft}) + (2)(15 \text{ ft}) + (3)(75 \text{ ft}) + (3)(75 \text{ ft})}{4 + 2 + 3 + 3}$$
$$= 45 \text{ ft}$$
$$\bar{y}_R = \frac{(4)(15 \text{ ft}) + (2)(75 \text{ ft}) + (3)(75 \text{ ft}) + (3)(15 \text{ ft})}{4 + 2 + 3 + 3}$$
$$= 40 \text{ ft}$$

## 94. Answer C

*SI solution*

Per the UBC [Sec. 1630.6], the accidental eccentricity should be calculated as 5% of the building dimension perpendicular to the direction of the applied load.

$$(L)(5\%) = 1.2 \text{ m}$$
$$L = \frac{1.2 \text{ m}}{0.05}$$
$$= 24.0 \text{ m}$$

*Customary U.S. solution*

Per the UBC [Sec. 1630.6], the accidental eccentricity should be calculated as 5% of the building dimension perpendicular to the direction of the applied load.

$$(L)(5\%) = 4 \text{ ft}$$
$$L = \frac{4 \text{ ft}}{0.05}$$
$$= 80 \text{ ft}$$

## 95. Answer C

*SI solution*

$$T_{\text{N-S}} = V(e_{\text{N-S}} + e_a)$$

The eccentricity is the distance between the centers of mass and rigidity, measured in a direction perpendicular to the lateral force.

$$e = \bar{x} - \bar{x}_R = 12.2 \text{ m} - 11.6 \text{ m}$$
$$= 0.6 \text{ m}$$

Per the UBC [Secs. 1630.6 and 1630.7], accidental eccentricity, $e_a$, is taken as 5% of the building dimension perpendicular to the lateral force.

$$e_a = (5\%)(24.4 \text{ m}) = 1.2 \text{ m}$$
$$T_{\text{N-S}} = V(e + e_a)$$
$$= (222\,410 \text{ N})(0.6 \text{ m} + 1.2 \text{ m})$$
$$= 400\,338 \text{ N·m})$$

*Customary U.S. solution*

$$T_{\text{N-S}} = V(e_{\text{N-S}} + e_a)$$

The eccentricity is the distance between the centers of mass and rigidity, measured in a direction perpendicular to the lateral force.

$$e = \bar{x} - \bar{x}_R = 40 \text{ ft} - 38 \text{ ft}$$
$$= 2 \text{ ft}$$

Per the UBC [Secs. 1630.6 and 1630.7], accidental eccentricity, $e_a$, is taken as 5% of the building dimension perpendicular to the lateral force.

$$e_a = (5\%)(80 \text{ ft}) = 4 \text{ ft}$$
$$T_{\text{N-S}} = V(e + e_a)$$
$$= (50{,}000 \text{ lbf})(2 \text{ ft} + 4 \text{ ft})$$
$$= 300{,}000 \text{ ft-lbf}$$

## 96. Answer A

The lateral force resultant acts through the center of mass, while the resisting force resultant acts through the center of rigidity. Thus, the building acts as though pinned at the center of rigidity, and a torsional moment is developed. For loading in the E-W direction, lateral force is acting through the center of mass and rotates around the center of rigidity.

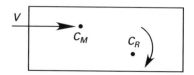

Therefore, for torsional moment the direction of rotation is clockwise.

## 97. Answer B

*SI solution*

$$e_x = 12.2 \text{ m} - 12.2 \text{ m}$$
$$= 0 \text{ m}$$
$$e_a = (5\%)(24.4 \text{ m})$$
$$= 1.2 \text{ m}$$
$$e = e_x + e_a = 0.0 \text{ m} + 1.2 \text{ m}$$
$$= 1.2 \text{ m}$$
$$T = Ve = (222\,410 \text{ N})(1.2 \text{ m})$$
$$= 266\,892 \text{ N·m}$$

The lateral force due to torsion is

$$F_t = T\left(\frac{Rd}{\Sigma Rd^2}\right)$$

$d$ is the distance from the center of rigidity perpendicular to the wall.

| wall | $R_x$ | $R_y$ | $d_x$ | $d_y$ | $Rd$ | $Rd^2$ |
|------|-------|-------|-------|-------|------|--------|
| N | 6 | – | – | +3.6 m | +21.6 m | 77.8 m$^2$ |
| S | 4 | – | – | –5.5 m | –22.0 m | 121.0 m$^2$ |
| E | – | 3 | +12.2 m | – | +36.6 m | 446.5 m$^2$ |
| W | – | 3 | –12.2 m | – | –36.6 m | 446.5 m$^2$ |

$$\Sigma Rd^2 = 1091.8 \text{ m}^2$$

For the north wall,

$$F_t = \frac{(266\,892 \text{ N·m})(21.6 \text{ m})}{1091.8 \text{ m}^2}$$
$$= 5280 \text{ N}$$

*Customary U.S. solution*

$$e_x = 40 \text{ ft} - 40 \text{ ft}$$
$$= 0 \text{ ft}$$
$$e_a = (5\%)(80 \text{ ft})$$
$$= 4 \text{ ft}$$
$$e = e_x + e_a = 0 \text{ ft} + 4 \text{ ft}$$
$$= 4 \text{ ft}$$
$$T = Ve = (50,000 \text{ lbf})(4 \text{ ft})$$
$$= 200,000 \text{ ft-lbf}$$

The lateral force due to torsion is

$$F_t = T\left(\frac{Rd}{\Sigma Rd^2}\right)$$

$d$ is the distance from the center of rigidity perpendicular to the wall.

| wall | $R_x$ | $R_y$ | $d_x$ | $d_y$ | $Rd$ | $Rd^2$ |
|------|-------|-------|-------|-------|------|--------|
| N | 6 | – | – | +12 ft | +72 ft | 864 ft$^2$ |
| S | 4 | – | – | –18 ft | –72 ft | 1296 ft$^2$ |
| E | – | 3 | +40 ft | – | +120 ft | 4800 ft$^2$ |
| W | – | 3 | –40 ft | – | –120 ft | 4800 ft$^2$ |

$$\Sigma Rd^2 = 11,760 \text{ ft}^2$$

For the north wall,

$$F_t = \frac{(200,000 \text{ ft-lbf})(72 \text{ ft})}{11,760 \text{ ft}^2}$$
$$= 1224 \text{ lbf}$$

98. Answer B

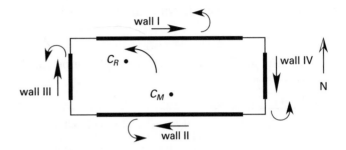

The building's center of rigidity fails to coincide with its center of mass, so the center of mass tends to rotate about the center of rigidity. As shown on the diaphragm, the direction of the torsional moment rotation is counterclockwise.

All walls contribute resistance to torsional moment. The direction of the torsional shear (resisting force) is opposite the direction of the torsional moment. Therefore, for walls I, II, III, and IV, the direction of the torsional shear is east, west, north, and south, respectively.

99. Answer B

*SI solution*

Consider the lateral forces due to shear and torsion.

The lateral force due to shear is

$$F_v = V\left(\frac{R}{\Sigma R}\right)$$
$$= (266\,900 \text{ N})\left(\frac{3}{3+1}\right)$$
$$= 200\,175 \text{ N}$$

The lateral force due to torsion is

$$e = 15.2 \text{ m} - 7.6 \text{ m}$$
$$= 7.6 \text{ m}$$
$$e_a = (5\%)(30.4 \text{ m})$$
$$= 1.5 \text{ m}$$

Note that the torsional moment rotation is counter-clockwise, or opposite to the direction of the applied lateral force. Therefore, the west wall receives only negative torsional components, and the minimum eccentricity must be used.

$$e_{\text{N-S}} = e - e_a = 7.6 \text{ m} - 1.5 \text{ m}$$
$$= 6.1 \text{ m}$$

$$T_{\text{N-S}} = Ve = (266\,900 \text{ N})(6.1 \text{ m}) \left( \frac{1 \text{ kN}}{1000 \text{ N}} \right)$$
$$= 1628 \text{ kN·m}$$

$d$ is the distance from the center of rigidity perpendicular to the wall.

| wall | $R_x$ | $R_y$ | $d_x$ | $d_y$ | $Rd$ | $Rd^2$ |
|------|-------|-------|-------|-------|------|--------|
| N | 3 | – | – | +7.6 m | +22.8 m | 173.3 m² |
| S | 5 | – | – | –4.6 m | –23.0 m | 105.8 m² |
| E | – | 1 | +22.8 m | – | +22.8 m | 519.8 m² |
| W | – | 3 | –7.6 m | – | –22.8 m | 173.3 m² |

$$\Sigma Rd^2 = 972.2 \text{ m}^2$$

$$F_t = T \left( \frac{Rd}{\Sigma Rd^2} \right)$$

$$= \frac{(1628 \text{ kN·m}) \left( 1000 \, \frac{\text{N}}{\text{kN}} \right)(-22.8 \text{ m})}{972.2 \text{ m}^2}$$

$$= -38\,180 \text{ N}$$

The total lateral force is the sum of the shear force and torsional force. Note that the torsional moment may not be neglected when negative. This was specified in the 1994 UBC, but is not specified in the 1997 UBC.

$$F_{\text{total}} = F_v + F_t$$
$$= 200\,175 \text{ N} - 38\,180 \text{ N}$$
$$= 161\,995 \text{ N}$$

*Customary U.S. solution*

Consider the lateral forces due to shear and torsion.

The lateral force due to shear is

$$F_v = V \left( \frac{R}{\Sigma R} \right)$$

$$= (60,000 \text{ lbf}) \left( \frac{3}{3+1} \right)$$

$$= 45,000 \text{ lbf}$$

The lateral force due to torsion is

$$e = 50 \text{ ft} - 25 \text{ ft}$$
$$= 25 \text{ ft}$$

$$e_a = (5\%)(100 \text{ ft})$$
$$= 5 \text{ ft}$$

Note that the torsional moment rotation is counter-clockwise, or opposite to the direction of the applied lateral force. Therefore, the west wall receives only negative torsional components and the minimum eccentricity must be used.

$$e_{\text{N-S}} = e - e_a = 25 \text{ ft} - 5 \text{ ft}$$
$$= 20 \text{ ft}$$

$$T_{\text{N-S}} = Ve = (60,000 \text{ lbf})(20 \text{ ft})$$
$$= 1200 \text{ ft-k}$$

$d$ is distance from the center of rigidity perpendicular to the wall.

| wall | $R_x$ | $R_y$ | $d_x$ | $d_y$ | $Rd$ | $Rd^2$ |
|------|-------|-------|-------|-------|------|--------|
| N | 3 | – | – | +25 ft | +75 ft | 1875 ft² |
| S | 5 | – | – | –15 ft | –75 ft | 1125 ft² |
| E | – | 1 | +75 ft | – | +75 ft | 5625 ft² |
| W | – | 3 | –25 ft | – | –75 ft | 1875 ft² |

$$\Sigma Rd^2 = 10{,}500 \text{ ft}^2$$

$$F_t = T \left( \frac{Rd}{\Sigma Rd^2} \right)$$

$$= \frac{(1200 \text{ ft-k}) \left( 1000 \, \frac{\text{lbf}}{\text{k}} \right)(3)(-25 \text{ ft})}{10{,}500 \text{ ft}^2}$$

$$= -8571 \text{ lbf}$$

The total lateral force is the sum of the shear force and torsional force. Note that the torsional moment may not be neglected when negative. This was specified in the 1994 UBC, but is not specified in the 1997 UBC.

$$F_{\text{total}} = F_v + F_t$$
$$= 45,000 \text{ lbf} - 8571 \text{ lbf}$$
$$= 36,429 \text{ lbf}$$

# CHAPTER 4
## DETAILS OF STRUCTURES

### TOPICS

Allowable Drift

Anchorage

Anchorage Force

Anchor Spacing

Axial Force

Base Isolation

Blocking

Braced Frame

Cantilever Frame Analysis

Cast-in-Place Concrete

Concrete Flexural Member

Concrete Slab

Connector

Cross-Grain Bending

Cross-Grain Tension

Design Yielding

Ductility

Fastener

Field Nailing Spacing

Interstory Deflection

Link Beam

Load Deformation Curve

Load Path

L-Shaped Building

Metal Strap

Nailing

Nail Spacing Requirement

Non-Reinforced Masonry

Plain Concrete

Plastic Hinge

Plywood Load Case

Plywood Shear Wall

Portal Frame Analysis

Primary Moment

Secondary Moment

Seismic Energy Absorption Capacity

Seismic Hoop/Tie

Shear Wall Action

Steel Reinforcement

Steel-Strain Diagram

Tension Lap Splice

Torsional Shear

Torsion Moment

Total Shear

Welding

Yielding

1. For wood-frame walls, the sheathing materials used to develop shear wall action include

   A. gypsum wallboards.
   B. plywood panels.
   C. fiberboards.
   D. all of the above.

2. In the design of wood structural panel shear wall, which of the following should be addressed?

   A. chord and strut design
   B. anchorage requirements
   C. nailing
   D. all of the above

3. A wood structural panel displays a span rating of "32/16 INCH." What do these numbers represent?

   A. the minimum recommended span for roofing (inches)/the minimum recommended span for subflooring (inches)
   B. the minimum recommended span for roofing (inches)/the maximum recommended span for subflooring (inches)
   C. the maximum recommended span for roofing (inches)/the minimum recommended span for subflooring (inches)
   D. the maximum recommended span for roofing (inches)/the maximum recommended span for subflooring (inches)

4. What are the standard dimensions of a roof wood structural panel?

   A. 3.5 ft × 7.4 ft  (1.0 m × 2.3 m)
   B. 4.0 ft × 4.0 ft  (1.2 m × 1.2 m)
   C. 4.0 ft × 6.0 ft  (1.2 m × 1.8 m)
   D. 4.0 ft × 8.0 ft  (1.2 m × 2.4 m)

5. The rafters and joists of a one-story wood structure are blocked. The main purpose of the blocking is to

   A. prevent drag strut force.
   B. prevent overturning force.
   C. prevent lateral buckling.
   D. decrease nailing.

6. For wood structures, what types of connectors or fasteners may be used?

   A. machine bolts
   B. nails
   C. lag bolts
   D. all of the above

7. Nails, lag bolts, and machine bolts in wood shall be oriented so that they are loaded in

   I. tension.
   II. shear.
   III. bending.

   A. I and II
   B. I and III
   C. II and III
   D. I only

8. For diaphragm design purposes, the framing and wood structural panel layout should be used to determine the load case type based on UBC Table 23-II-H. These load cases essentially depend on

   A. the direction of the lateral load on diaphragm.
   B. the direction of the continuous panel joint.
   C. the direction of the unblocked edge.
   D. both A and B.

9. The UBC load cases (Table 23-II-H) for horizontal diaphragm design do not depend on

   A. the direction of the lateral load on the diaphragm.
   B. the direction of the continuous panel joint.
   C. the direction of the wood structural panel.
   D. any of the above.

10. Refer to UBC Table 23-II-H for horizontal diaphragms. When the lateral load is parallel to the continuous panel joints, what are the load case numbers?

   I. Cases 1 and 2
   II. Cases 3 and 4
   III. Cases 5 and 6

   A. I and II
   B. II and III
   C. III only
   D. I, II, and III

11. Refer to UBC Table 23-II-H for horizontal diaphragms. When the lateral load is perpendicular to continuous panel joints, what are the load case numbers?

    I. Cases 1 and 2
    II. Cases 3 and 4
    III. Cases 5 and 6

    A. I only
    B. I and II
    C. I and III
    D. II and III

12. For developing the design capacity of a nail, what should the minimum distance be from the edge of a wood structural panel to the center of the nail?

    A. $\frac{3}{8}$ in (10 mm)
    B. $\frac{1}{2}$ in (13 mm)
    C. $\frac{3}{4}$ in (19 mm)
    D. 1 in (25 mm)

13. For wood structural panel roof diaphragms, what does the field nailing spacing depend on?

    A. the direction of loading
    B. the blocked or unblocked diaphragm
    C. the size of framing members
    D. none of the above

14. What is the standard length of a 10d box nail?

    A. 2 in (51 mm)
    B. $2\frac{1}{2}$ in (64 mm)
    C. 3 in (76 mm)
    D. $3\frac{1}{2}$ in (89 mm)

15. 50d common nails are driven perpendicular to the grain of the wood. What is the minimum required penetration length specified by the UBC?

    A. 1.22 in (31 mm)
    B. 1.95 in (50 mm)
    C. 2.68 in (68 mm)
    D. 3.42 in (87 mm)

16. A wood diaphragm is used to laterally support concrete shear walls of a one-story building in Los Angeles. Anchorage may be accomplished by the use of

    I. toenails.
    II. connections.
    III. nailing subject to withdrawal.

    A. I only
    B. II only
    C. I and II
    D. II and III

17. A wood diaphragm is used to laterally support the concrete shear walls of a one-story warehouse in seismic zone 3. Wood ledgers or framing should be used in

    I. cross-grain bending.
    II. cross-grain tension.

    A. I only
    B. II only
    C. I and II
    D. neither I nor II

18. Plain concrete walls should not be used in which of the following seismic zones?

    I. 0 and 1
    II. 2A and 2B
    III. 3 and 4

    A. I only
    B. III only
    C. II and III
    D. Plain concrete walls should not be used in any zone.

19. What should be the minimum thickness of cast-in-place toppings over precast roof and floor diaphragms?

    A. $1\frac{1}{2}$ in (38 mm)
    B. $2\frac{1}{2}$ in (64 mm)
    C. 3 in (76 mm)
    D. $3\frac{1}{2}$ in (89 mm)

20. The minimum thickness of concrete floor slabs supported directly on the ground should be

    A. 2 in (51 mm).
    B. $2\frac{1}{2}$ in (64 mm).
    C. $3\frac{1}{2}$ in (89 mm).
    D. 4 in (102 mm).

21. For achieving ductility in concrete frames, which of the following should be ensured?

    A. design procedure by the "ultimate strength design method"
    B. tensile reinforcement yields before the concrete
    C. both (A) and (B)
    D. none of the above

22. When reinforcement of concrete slabs is required, where should the steel reinforcing generally be placed?

    A. near the top of the slab
    B. where the concrete is expected to go into tension
    C. near the bottom of the slab
    D. at the middle of the slab

23. For reinforced concrete structures, what should be the prime concern in the design of joints where the elements of a structure intersect each other?

    A. the location of the anchorage of reinforcement
    B. the location of the transfer of axial load
    C. the location of the shear
    D. all of the above

24. As shear reinforcement, what would you use to confine concrete in a column?

    A. hoops/ties
    B. stirrups
    C. hooks
    D. crossties

25. For a ductile concrete moment-resisting space frame, where should design yield hinging occur?

    A. in beams only
    B. in columns only
    C. in both beams and columns
    D. it is not needed

26. You are to design the seismic hoops for concrete flexural members. Where would you use them?

    A. when yielding is expected
    B. at the ends of built-in beams
    C. at plastic hinges
    D. for all of the above

27. Based on the UBC, lap splices of flexural reinforcement of the concrete flexural members should be used

    A. within the joints.
    B. within a distance of twice the member depth from the face of joint.
    C. where flexural yielding is expected.
    D. for none of the above.

28. For a concrete flexural member with required reinforcement in seismic zones 3 and 4, lap splices are permitted by the UBC

    A. where yielding is expected.
    B. within beam-column joints.
    C. at plastic hinge points.
    D. within the center half of the member length.

29. For concrete collector elements, which of the following types of steel reinforcing is not permitted?

    A. bars larger than No. 6
    B. bars less than No. 9
    C. plain bars
    D. low alloy A706 steel bars

30. Based on UBC Chapter 19 requirements, the reinforcement bars (including ties, stirrups, and spirals) placed in cast-in-place concrete beams and columns not exposed to weather should be covered with a thickness of concrete not less than

    A. $\frac{3}{4}$ in (19 mm).
    B. $1\frac{1}{2}$ in (38 mm).
    C. 2 in (51 mm).
    D. 3 in (76 mm).

31. Non-reinforced masonry walls are prohibited in which of the seismic zones?

    A. in zone 4 only
    B. in zones 3 and 4 only
    C. in zones 2, 3, and 4
    D. in all zones

32. Based on UBC requirements, masonry shear walls should be designed to resist what percentage of the base shear in seismic zones 3 and 4?

    A. 50%
    B. 85%
    C. 100%
    D. 150%

33. Based on the UBC, what should be the maximum allowable shear stress for masonry walls when shear reinforcement is designed to take all the shear?

    A.  23 lbf/in$^2$  (159 kPa)
    B.  35 lbf/in$^2$  (241 kPa)
    C.  75 lbf/in$^2$  (517 kPa)
    D.  87 lbf/in$^2$  (600 kPa)

34. Concrete and masonry walls should be anchored to all floors, roofs, and other structural elements that provide lateral support for the walls. In seismic zone 3, per UBC requirements, that anchorage should be capable of resisting what minimum horizontal force per lineal foot (meter) of wall?

    A.  100 lbf/ft  (1460 N/m)
    B.  150 lbf/ft  (2190 N/m)
    C.  280 lbf/ft  (4090 N/m)
    D.  $0.5C_aI$ times the dead plus live load

35. Per UBC requirements, concrete or masonry walls should be anchored to all floors, roofs, and other structural elements that provide lateral support for the walls. For walls that are not designed to resist bending between anchors, what should be the maximum anchor spacing?

    A.  2 ft  (0.6 m)
    B.  4 ft  (1.2 m)
    C.  6 ft  (1.8 m)
    D.  8 ft  (2.4 m)

36. When quality assurance provisions do not include the requirements of any testing during construction, the allowable design masonry stress as stated in the UBC should be

    A.  reduced by $\frac{1}{2}$.
    B.  increased by $\frac{1}{2}$.
    C.  reduced by $\frac{2}{3}$.
    D.  increased by $\frac{2}{3}$.

37. The following illustration represents a typical stress-strain structural diagram for which of the following?

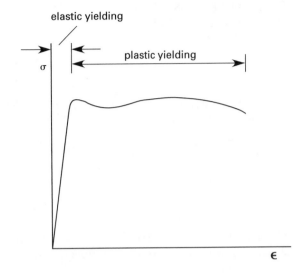

    A.  concrete
    B.  reinforced concrete
    C.  steel
    D.  masonry

38. For a well-designed and well-constructed steel structure, choose the correct statement.

    A.  Plastic hinges form where the moments are zero.
    B.  Plastic hinges form at maximum moment.
    C.  Steel columns of the structure never buckle.
    D.  Steel structures have a greater damping ratio compared with concrete structures.

39. For the following building shown in elevation, the forces on the wall are as illustrated. What is the axial force on the column due to the lateral loads?

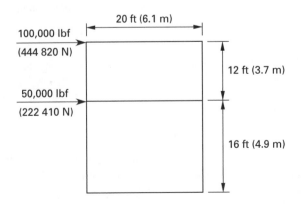

A. 75,000 lbf (333 740 N)
B. 100,000 lbf (447 650 N)
C. 130,000 lbf (581 950 N)
D. 180,000 lbf (805 780 N)

40. Among the following types of construction, which can best tolerate tension?

A. concrete
B. steel
C. masonry
D. wood

41. The load-deformation curves of different structures are shown. Identify the curve that represents the most ductile structure.

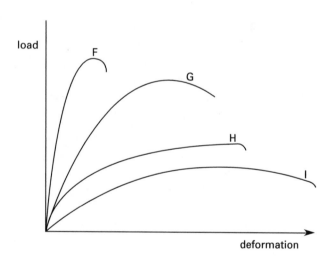

A. F
B. G
C. H
D. I

42. A nonuniform, L-shaped building is shown. At what point is the local stress concentration located?

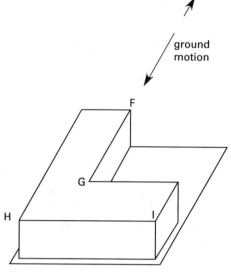

A. F
B. G
C. H
D. I

43. Which of the following braced frames has the best capability of yielding and absorbing seismic energy?

A. eccentric
B. diagonal
C. chevron
D. K

44. In an eccentrically braced frame, which of the following functions is performed by a link beam?

I. yielding in a severe earthquake
II. absorbing great amounts of seismic energy
III. transferring buckling to the other bracing members

A. I only
B. I and II
C. I and III
D. II and III

45. When welding along the edges of connecting plates with different thicknesses, the minimum weld size depends on the

I. thickness of the thicker plate.
II. thickness of the thinner plate.
III. length of plates.

A. I only
B. II only
C. I and II
D. I and III

For Problems 46 and 47, refer to the following set of assumptions.

The building shown should be designed to resist the effects of seismic ground motions based on UBC Chapter 16. The building is proposed to have a different type of structural system at each floor. The $R$ values for each floor are given in the illustration.

| | | |
|---|---|---|
| R = 8.5 | | 8th floor |
| R = 8.5 | | 7th floor |
| R = 4.5 | | 6th floor |
| R = 8.5 | | 5th floor |
| R = 8.5 | | 4th floor |
| R = 4.5 | | 3rd floor |
| R = 8.5 | | 2nd floor |
| R = 8.5 | | 1st floor |

46. In seismic load calculations, what value of $R$ would you use for the second and seventh floors, respectively?

A. 4.5 and 4.5
B. 4.5 and 8.5
C. 8.5 and 4.5
D. 8.5 and 8.5

47. In seismic load calculations, what value of $R$ would you use for the first and fourth floors, respectively?

A. 8.5 and 8.5
B. 8.5 and 4.5
C. 4.5 and 4.5
D. 8.5 and 13.0

48. Which of the following statements is correct about a structure fitted with base isolation?

A. The building period will decrease.
B. The building damping ratio will decrease.
C. The building acceleration will decrease.
D. The building drift will increase.

49. A structure is generally suitable for seismic base isolation where

   I. the site (soil) does not produce a long period.
   II. the soil is soft.
   III. the structure is rigidly anchored to its foundation.

A. I only
B. I and II only
C. I and III only
D. I, II, and III

50. The elastic design level response displacement, $\Delta_S$, for a special moment-resisting steel structure with 15 ft (4.6 m) story height is computed to be 0.2 in (5 mm). Architectural items are attached to the structure by mechanical connections and fasteners. In case of an earthquake, what should be the maximum relative movement between stories allowed by these connections for this structure to be free of possible architectural damage?

A. 0.3 in (7.5 mm)
B. 0.5 in (12.7 mm)   $\Delta_m = (0.7)R\Delta_s$
C. 0.9 in (23 mm)
D. 1.2 in (30 mm)

51. Force $F$ is acting on the frame shown. The framing system is equally spaced between columns. Applying the portal method of frame analysis, what should the distributed shear be at column II?

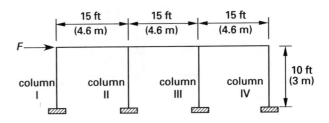

A. $\frac{1}{8}F$
B. $\frac{1}{6}F$
C. $\frac{1}{4}F$
D. $\frac{1}{3}F$

52. In the consideration of seismic and wind forces, building frames should be analyzed by which of the following methods of analysis?

    A. by portal frame analysis
    B. by cantilever frame analysis
    C. by all of the above
    D. by none of the above

53. Identify the principal assumption that is the basis for the cantilever analysis of frames subject to lateral forces.

    A. All horizontal planes remain horizontal.
    B. The entire floor remains plane as one unit.
    C. All interior columns carry the same shear.
    D. None of the above are true.

54. Identify the method of frame analysis that assumes that all horizontal planes remain horizontal under lateral forces.

    A. cantiliver frame analysis
    B. factor frame analysis
    C. space frame analysis
    (D) portal frame analysis

For Problems 55 and 56, refer to the given set of assumptions.

The column members of a one-story building are loaded as shown in the following illustration. The allowable drift is limited to 0.5% of the height. The building is in seismic zone 4.

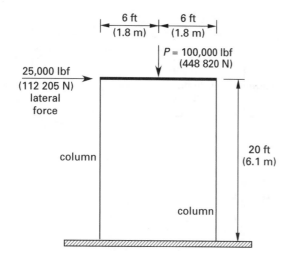

55. Determine the magnitude of the primary moment.

    A. 10,000 ft-lbf   (13 690 N·m)
    B. 100,000 ft-lbf   (136 890 N·m)
    C. 500,000 ft-lbf   (684 450 N·m)
    D. 600,000 ft-lbf   (821 340 N·m)

56. Determine the magnitude of the secondary moment.

    A. 10,000 ft-lbf   (13 460 N·m)
    B. 100,000 ft-lbf   (134 580 N·m)
    C. 500,000 ft-lbf   (672 910 N·m)
    D. 600,000 ft-lbf   (807 490 N·m)

For Problems 57 and 58, refer to the following set of assumptions.

The following figure shows a typical connection between a wood diaphragm and a concrete or masonry wall.

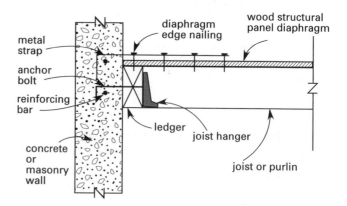

57. The purpose of the anchor bolt is to transfer what loads to the concrete or masonry walls?

    A. the vertical load from floor or roof framing
    B. the lateral load parallel to the wall
    C. both (A) and (B)
    D. none of the above

58. For transferring the lateral force from a wood structural panel diaphragm to a masonry wall and through the wall to the foundation, the path of the lateral load through the following components should be in what order?

    I. the edge nails
    II. the ledger
    III. the anchor bolts

MOMENTS

A. from I to III
B. from II to III
C. through III only
D. from I to II to III

For Problems 59 and 60, refer to the following set of assumptions.

The figure shows a typical connection between a wood diaphragm and a concrete or masonry wall.

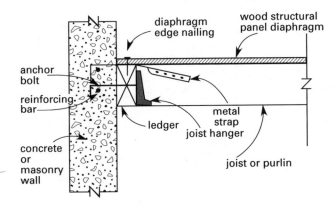

59. The metal strap (wall tie) should be nailed or bolted to

A. wood joists.
B. added blocking.
C. both (A) and (B).
D. none of the above.

60. Metal straps

A. provide a tension connection between two members.
B. resist the diaphragm lateral force.
C. provide blocking.
D. are not needed at all.

For Problems 61 and 62, refer to the following set of assumptions.

The following figure shows a typical connection between a horizontal diaphragm and a wall. The diaphragm is a steel deck with concrete fill, and the wall is concrete or masonry.

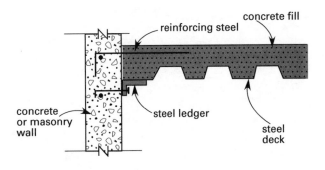

61. The horizontal reinforcing steel shown in the concrete diaphragm is used to

A. resist vertical loads.
B. provide compression reinforcement.
C. transfer buckling.
D. transfer diaphragm shear.

62. What force determines the spacing and size of the reinforcement dowels?

A. anchorage force
B. inertia force
C. strut force
D. none of the above

# CHAPTER 4

## Solutions

**1. Answer D**
Shear walls function as shear-resisting elements in carrying lateral loads. Sheathing materials suitable for producing shear wall action include plywood sheathing, gypsum fiberboards, particleboard, Portland cement plaster, and hardboard panels.

**2. Answer D**
In designing wood structural panel shear walls (and also horizontal diaphragms), it is necessary to account for horizontal and vertical diaphragm shears, chords, struts, diaphragm ratios, blocking, and connectors (i.e., nails, bolts, and ties).

**3. Answer D**
The span rating specifies the precalculated suitability of wood structural panels for roofing and subflooring. In the 32/16 span rating, the 32 represents the maximum recommended span for roofing (inches), and the 16 specifies the maximum recommended span for subflooring (inches).

**4. Answer D**
For roof, floor, and wall sheathing design applications, the standard size of a wood structural panel is 4 ft (1.2 m) by 8 ft (2.4 m). The UBC [Sec. 2315.3.3] gives wood structural panel diaphragms provisions. For seismic zones 3 and 4 refer to the UBC [Sec. 2315.5.3].

**5. Answer C**
Per UBC [Sec. 2320.12.8], roof rafters, ceiling joists, and beams should be prevented from rotation and lateral displacement. The purpose of blocking is to produce lateral support and prevent joists from buckling.

**6. Answer D**
Wood connectors and fasteners are used to carry forces between wood members and between wood and metal/ concrete members. The connectors may be nails and spikes, wood screws, lag bolts, joist hangers and framing anchors, and metal plate connectors. UBC Chapter 23, Division III, Part III, gives allowable loads and provisions for the installation of connectors and fasteners.

**7. Answer C**
Wood connectors and fasteners are used to transmit forces between wood members and between wood and metal/concrete members. They should be laterally loaded and should not be subjected to a greater load in shear and bending than the safe lateral strength as set forth in UBC Chapter 23, Division III, Part III. It is desirable to avoid loading fasteners in tension (i.e., wood connectors and fasteners driven perpendicular to the direction of the grain) since the ultimate strength in tension is extremely low.

**8. Answer D**
UBC Table 23-II-H illustrates the types of wood structural panel diaphragm layouts. These layouts are referred to as *load cases*. The direction of the lateral load on diaphragms and the direction of the continuous panel joints determine the type of load case.

**9. Answer C**
The direction of the wood structural panel does not determine the type of load case. Wood structural panels can be installed horizontally, vertically, or diagonally (at an angle of 45° to the supports). The orientation of the continuous panel joint, not the orientation of the panel itself, determines the load case.

**10. Answer B**
Use Cases 3, 4, 5, and 6 of the diaphragm panel layout configurations in UBC Table 23-II-H. This is because the continuous panel joints are parallel to the direction of lateral load. For Cases 1 and 2, the continuous panel joints are perpendicular to the direction of lateral load.

**11. Answer C**

Use Cases 3, 4, 5, and 6 of the diaphragm panel layout configurations in UBC Table 23-II-H. This is because the continuous panel joints are perpendicular to the direction of lateral load. For Cases 3 and 4, the continuous panel joints are parallel to the direction of lateral load.

**12. Answer A**

The nail spacing differs at various points in the diaphragm. The minimum distance from the edge of a plywood panel to the center of a nail is $\frac{3}{8}$ in (10 mm). This distance mainly prevents the nail head from pulling away through the edge of the wood structural panel. Refer to UBC [Sec. 2315.3.3] for more details.

**13. Answer D**

*Field nailing* refers to the nailing of wood structural panel diaphragms to intermediate framing members. UBC provisions specify the field nailing spacing to be 12 in (30 cm) O.C. for roofs and 10 in O.C. for floors.

**14. Answer C**

UBC Tables 23-III-C-1 and 23-III-C-2 give lateral design values for box and common nails. The standard length of 10d box nails is 3 in (76 mm).

**15. Answer C**

Based on the UBC [Sec. 2318.3.3] and UBC Tables 23-III-C-1 and 23-III-C-2, the minimum required penetration length for 50d common nail is 11 times its diameter. The diameter of 50d nail is 0.244 in (6.2 mm).

**16. Answer B**

In seismic zones 2, 3, and 4, when wood diaphragms are used to laterally support concrete or masonry walls, the UBC [Sec. 1633.2.9, Item 5] specifies that anchorage should not be achieved by applying toenails or nails subject to withdrawal.

**17. Answer D**

The UBC [Sec. 1633.2.9, Item 5] specifies that in seismic zones 2, 3, and 4, when wood diaphragms are used to laterally support concrete or masonry walls, wood ledgers or framing should not be used in cross-grain bending or cross-grain tension. To prevent these errors, additional tension connection straps must be used to anchor the diaphragm to the ledger or framing.

**18. Answer C**

*Plain concrete* is concrete that is either unreinforced or inadequately reinforced compared to the minimum amount specified in the UBC for reinforced concrete. Based on the UBC [Secs. 1922.2.2, 1922.2.5, and 1922.10.3], plain concrete walls should not be used in

seismic zones 2, 3, and 4. The UBC [Sec. 1922.10.3] lists exceptions where structural plain concrete is permitted. Refer to that section for more details.

**19. Answer C**

A cast-in-place topping on a precast floor system functions as the diaphragm. Its thickness, based on the UBC [Sec. 1921.6.12, Item 3], should not be less than 3 in (76 mm).

**20. Answer C**

For concrete floor slabs supported directly on the ground, the UBC [Sec. 1900.4.4], mandates a minimum thickness of $3\frac{1}{2}$ inches (89 mm).

**21. Answer C**

Concrete members should be designed to have adequate stiffness to limit deflections or deformations that may adversely influence the strength of the structure. Therefore, to achieve ductility, the design procedure should be by the "ultimate strength design method," and design strength for reinforcement should be based on the yield strength of reinforcement.

**22. Answer B**

Reinforced slabs resist flexural stresses. The steel reinforcing should be generally placed where the concrete is expected to go into tension.

**23. Answer D**

In designing ductile concrete frame joints, the location of the anchorage of reinforcement, transfer of axial load, and shear are critical. Designing with deficiencies in these locations, due to cracking or crushing of the core, causes the concrete members to withdraw and disintegrate toward a state of collapse.

**24. Answer A**

Ties/hoops (confinements) are loops of reinforcing bars enclosing longitudinal reinforcements. They apply to lateral reinforcement in compression members (columns). Provisions are stated in the UBC [Sec. 1907.10.5].

**25. Answer A**

Design yield hinging should occur in beams only; special reinforcement is required at points where the yielding is expected.

26. Answer D
Seismic hoops apply to lateral reinforcement in flexural members. They should be placed where yielding is expected. Yielding occurs at the ends of built-in beams and at points where the moments are greatest (i.e., plastic hinges).

27. Answer D
According to the UBC [Sec. 1921.3.2.3], lap splices of flexural reinforcement should be allowed only if hoop or spiral reinforcement is furnished over the lap length. "Lap splices shall not be used (1) within the joints, (2) within a distance of twice the member depth from the face of joint, and (3) at locations where analysis indicates flexural yielding caused by inelastic lateral displacements of the frame."

28. Answer D
Based on the UBC [Sec. 1921.3.2.3], lap splices of flexural reinforcement can be used within the center half of the member length but not less than twice the member depth from the face of the joint.

29. Answer C
Based on the UBC [Sec. 1902], plain reinforcement is reinforcement that does not conform to definition of deformed reinforcement.

According to the UBC [Sec. 1921.2.5.1], plain bars cannot be used for steel reinforcing. In frame members and in wall boundary elements, low alloy A706 reinforcement bars are required. Based on the UBC [Sec. 1921.2.5.2], billet steel A615 Grades 40 and 60 reinforcement may be allowed if certain conditions are met.

30. Answer B
According to the UBC [Sec. 1907.7.1, Item 3], cast-in-place concrete (nonprestressed) should provide a minimum of $1\frac{1}{2}$ in (38 mm) of concrete cover to protect the reinforcement.

31. Answer C
The UBC [Secs. 2106.1.12.3, Items 2 and 3, and 2106.1.12.4, Item 2.3] specifies that in seismic zones 2, 3, and 4, all masonry structures should be reinforced with both vertical and horizontal reinforcement.

32. Answer D
According to the UBC [Sec. 2107.1.7], in seismic zones 3 and 4, shear walls that resist seismic shear forces should be designed to resist 1.5 times the base shear.

33. Answer C
Assuming that quality assurance provisions include requirements for special inspection, based on the UBC [Sec. 2107.2.9], the maximum allowable stresses in masonry when shear reinforcement is designed to take all the shear is 75 lbf/in$^2$ (520 kPa).

34. Answer C
Per the UBC [Secs. 1605.2.3 and 1611.4], the minimum horizontal force for which anchorage should be designed is 280 pounds per lineal foot (4.09 kN/m) of wall. This provision provides the minimum adequate anchorage between the walls and the diaphragm. Note that per the UBC [Sec. 1633.2.8.1, Item 1], in seismic zone 4, the minimum specified anchorage force is 420 pounds per lineal foot (6.1 kN/m) of wall.

35. Answer B
According to the UBC [Sec. 1605.2.3], where anchor spacing exceeds 4 ft (1.2 m), walls should be designed to resist bending between anchors. The UBC [Secs. 1632, 1633.2.8, and 1633.2.9] give provisions for earthquake design requirements.

36. Answer A
The UBC [Sec. 2107.2.9], gives the allowable stresses in masonry. However, according to the UBC [Sec. 2107.1.2], these values are reduced by $\frac{1}{2}$ if the quality assurance provisions exclude requirements for special inspection as prescribed in the UBC [Sec. 1701].

37. Answer C
The curve schematically illustrates the variation of the critical load when an instability occurs. The elastic yielding range (i.e., flexibility) and plastic yielding range (i.e., ductility) are shown in the diagram. Based on the type of behavior shown in the illustration (maximum stress prior to yield stress level, and stress at a constant strain rate after initial yield), this stress-strain diagram is representative of a mild structural steel.

38. Answer B
Plastic behavior occurs after the plastic moment is reached. Therefore, for steel structures, plastic hinges are anticipated at points where the moments are greatest.

**39. Answer D**

*SI solution*

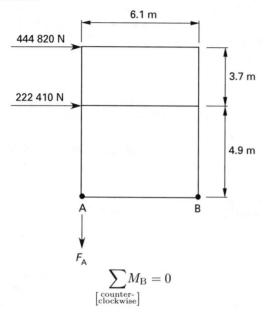

$$\sum M_{\mathrm{B}} = 0 \quad \left[\begin{smallmatrix}\text{counter-}\\\text{clockwise}\end{smallmatrix}\right]$$

$$F_{\mathrm{A}}(6.1 \text{ m}) - (444\,820 \text{ N})(3.7 \text{ m} + 4.9 \text{ m})$$
$$- (222\,410 \text{ N})(4.9 \text{ m})$$
$$= 0$$
$$F_{\mathrm{A}} = 805\,780 \text{ N}$$

*Customary U.S. solution*

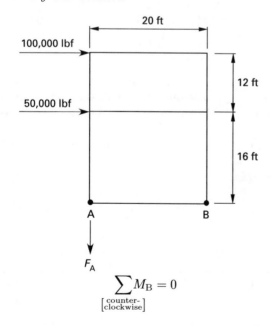

$$\sum M_{\mathrm{B}} = 0 \quad \left[\begin{smallmatrix}\text{counter-}\\\text{clockwise}\end{smallmatrix}\right]$$

$$F_{\mathrm{A}}(20 \text{ ft}) - (100,000 \text{ lbf})(12 \text{ ft} + 16 \text{ ft})$$
$$- (50,000 \text{ lbf})(16 \text{ ft})$$
$$= 0$$
$$F_{\mathrm{A}} = 180,000 \text{ lbf}$$

**40. Answer B**

Steel structures can best tolerate tension due to the properties of steel, such as its modulus of elasticity, yield point, yield-stress level, and so on.

**41. Answer D**

*Ductility* is the ability of a material to deform or yield without undergoing some sort of failure. It is usually represented as a ratio of some property at failure to that same property at the yield point. The curves shown represent the deformation capacities of the structures F, G, H, and I. Structure I, while having the lowest peak load capacity, is able to deform the most without a loss in capacity. Therefore it is the most ductile.

**42. Answer B**

In nonuniform L-shaped buildings, local stresses develop and concentrate through or into the heel of the L (i.e., at point G).

**43. Answer A**

Structures with eccentric braced frames undergo small drifts in low to modest earthquakes and perform as braced frames rather than as moment frames. In large earthquakes, special sections of the girders are designed to yield, making these structures highly ductile ($R = 7.0$). Ductile structures can absorb a great amount of seismic energy.

**44. Answer B**

An *eccentric braced frame* (EBF), as defined by the UBC [Sec. 2211.3], is a diagonal braced frame where at least one end of each bracing member is attached to a beam a short distance from a beam-to-column connection or from another beam-to-brace connection. The part of a beam between the brace end and the column is known as the *link beam*. Its function is to yield in shear and/or bending, thus preventing the buckling of the bracing members. Therefore, eccentric braced frames can absorb great amounts of seismic energy. The EBF should meet the requirements of the UBC [Sec. 2211.4, Item 10].

**45. Answer A**

The plates should be welded across the plate width, top and bottom, with fillet welds. For the parts that are joined, the minimum weld size to carry the required load depends on the thickness of the thicker plate.

**46. Answer B**

According to the UBC [Sec. 1630.4.2], the value of $R$ for the design of any story should be less than or equal to the value of $R$ used for the story above in the given direction. For the third floor, the value of $R$ is 4.5; thus $R$ for the second floor must be less than or equal to 4.5. For the eighth floor, the value of $R$ is 8.5; thus $R$ for the seventh floor must be less than or equal to 8.5.

**47. Answer A**

According to the UBC [Sec. 1630.4.2], in the design of any story, the value of $R$ should be less than or equal to the value of $R$ used for the story above. For the second floor, the value of $R$ is 8.5; thus $R$ for first floor must be less than or equal to 8.5. For the fifth floor, the value of $R$ is 8.5; thus $R$ for the fourth floor must be less than or equal to 8.5.

**48. Answer C**

Some structures (buildings and bridges) are not rigidly anchored to their foundations, but instead are equipped with base isolators. In an earthquake, the ground moves independently of buildings that are base isolated. As a result, the acceleration of these structures will be less than that for structures that do not have base isolation. The isolation system should be designed and constructed in accordance with the requirements of UBC Div. IV—Earthquake Regulations for Seismic-Isolated Structures—and the applicable requirements of UBC Chap. 16, Part IV.

**49. Answer A**

For structures fitted with base isolation, the acceleration decreases while the structure's natural period increases. Therefore, base isolation is not suitable where the soil is soft because the long period of soft soil may coincide with the increased period of the structure and produce resonance. A structure that is rigidly anchored to its foundation is, by definition, not base isolated. For the UBC definition of "isolation system," refer to Sec. 1655, Div. IV (Appendix Chap. 16).

**50. Answer B D**

*SI solution*

According to the UBC [Sec. 1633.2.4.2], elements that are attached to the exterior should accommodate movements of the structure based on $\Delta_M$. Such elements are supported by mechanical connections and fasteners as given in the problem statement. Note that exterior nonbearing, nonshear wall panels can be supported by means of cast-in-place concrete.

Based on the UBC [Sec. 1633.2.4.2, Item 1], these connections should allow for a relative movement between stories of not less than two times the story drift caused by wind, the calculated story drift, based on $\Delta_M$ or 12.7 mm, whichever is greater.

Per the UBC [Sec. 1630.9.2], the maximum inelastic response displacement, $\Delta_M$, should be computed from UBC Formula 30-17, $\Delta_M = 0.7R\Delta_S$.

From UBC Table 16-N, for a special moment-resisting steel structure, the value of $R$ is 8.5.

Therefore,

$$\Delta_M = (0.7)(8.5)(5 \text{ mm})$$
$$= 30 \text{ mm}$$

The displacement of 30 mm is greater than 12.7 mm. Thus, the connections should allow for a relative movement of 30 mm between stories.

*Customary U.S. solution*

According to the UBC [Sec. 1633.2.4.2], elements that are attached to the exterior should accommodate movements of the structure based on $\Delta_M$. Such elements are supported by mechanical connections and fasteners as given in the problem statement. Note that exterior nonbearing, nonshear wall panels can be supported by means of cast-in-place concrete.

Based on the UBC [Sec. 1633.2.4.2, Item 1], these connections should allow for a relative movement between stories of not less than two times the story drift caused by wind, the calculated story drift, based on $\Delta_M$ or $\frac{1}{2}$ in, whichever is greater.

Per the UBC [Sec. 1630.9.2], the maximum inelastic response displacement, $\Delta_M$, should be computed from UBC Formula 30-17, $\Delta_M = 0.7R\Delta_S$.

From UBC Table 16-N, for a special moment-resisting steel structure, the value of $R$ is 8.5.

Therefore,

$$\Delta_M = (0.7)(8.5)(0.2 \text{ in})$$
$$= 1.2 \text{ in}$$

The displacement of 1.2 in is greater than 0.5 in. Thus, the connections should allow for a relative movement of 1.2 in between stories.

**51. Answer D**
The portal method assumes that exterior columns carry half the shear of interior columns and that the interior columns carry equal shear.

$$V_I = 0.5V_{II}$$
$$V_{IV} = 0.5V_{II}$$
$$V_{II} = V_{III}$$

Thus,

$$F = V_I + V_{II} + V_{III} + V_{IV}$$
$$= 0.5V_{II} + V_{II} + V_{II} + 0.5V_{II}$$
$$= 3V_{II}$$
$$V_{II} = \tfrac{1}{3}F$$

**52. Answer C**
Rigid frame analysis can be performed by either the portal method or the cantilever method. These methods distribute the lateral forces to the resisting elements.

**53. Answer B**
The cantilever method assumes that the entire structure functions as a cantilever beam (i.e., the floors remain plane).

**54. Answer D**
As opposed to the cantilever method, the portal method assumes that horizontal diaphragms remain horizontal when subjected to lateral forces.

**55. Answer C**

*SI solution*

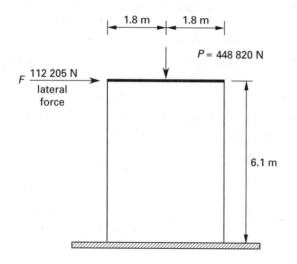

The primary moment is the moment due to the lateral forces caused by the earthquake.

For the primary moment,

$$Fh = (112\,205 \text{ N})(6.1 \text{ m})$$
$$= 684\,450 \text{ N·m}$$

*Customary U.S. solution*

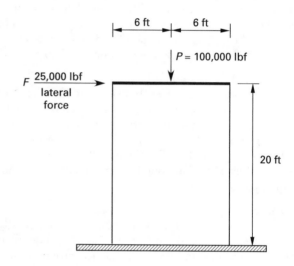

The primary moment is the moment due to the lateral forces caused by the earthquake.

For the primary moment,

$$Fh = (25,000 \text{ lbf})(20 \text{ ft})$$
$$= 500,000 \text{ ft-lbf}$$

**56. Answer A**

*SI solution*

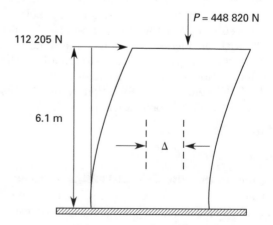

The secondary moment is the additional column stress caused by the $P$-$\Delta$ effect.

$$\Delta = (0.5\%)(6.1 \text{ m}) \left(1000 \; \frac{\text{mm}}{\text{m}}\right)$$
$$= 30 \text{ mm}$$

For the secondary moment,

$$P\text{-}\Delta = (448\,820 \text{ N})(30 \text{ mm}) \left(0.001 \; \frac{\text{m}}{\text{mm}}\right)$$
$$= 13\,460 \text{ N·m}$$

*Customary U.S. solution*

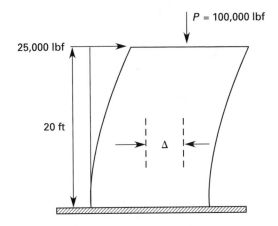

The secondary moment is the additional column stress caused by the $P$-$\Delta$ effect.

$$\Delta = (0.5\%)(20 \text{ ft}) = 0.1 \text{ ft}$$

For the secondary moment,

$$P\text{-}\Delta = (100,000 \text{ lbf})(0.1)$$
$$= 10,000 \text{ ft-lbf}$$

**57. Answer C**
According to the UBC [Sec. 1605.2.3], anchor bolts between the diaphragms and the vertical resisting elements are designed to resist lateral loads parallel to the resisting elements. They also provide support for the vertical loads.

**58. Answer D**
Diaphragms transfer lateral forces to the shear walls through the connections. Connections between all elements should be capable of transmitting the applied force from one element to another. Therefore, the path of the lateral forces is from the edge nails to the ledger, and then to the anchor bolts. Shear walls carry the forces transferred from the diaphragms. Also, in addition to their own inertia forces, shear walls carry the forces transferred from the diaphragms to the foundations.

**59. Answer C**
All parts of a structure should be interconnected by connections (e.g., metal straps) that are capable of transmitting the lateral force produced by other connected parts. The anchors are embedded in the wall and are nailed or bolted to the joists and added blocking.

**60. Answer A**
Metal straps are diaphragm-to-wall anchors. They tie concrete or masonry walls to horizontal diaphragms by providing a tension connection between two members.

**61. Answer D**
In resisting the designed lateral and vertical loads, the anchorages between the walls and roof enable the building to function as a unit. Connections between diaphragms and walls (e.g., reinforcing steel) should be designed to transfer the horizontal forces to the resisting elements.

**62. Answer A**
The design lateral forces should be used to design members and connections. The transfer of shear to the walls is achieved by the use of reinforcing steel connections. Therefore, the anchorage force of the wall to the diaphragm determines the spacing and size of the reinforcing steel.

# CHAPTER 5
## DESIGN PROBLEMS

### CONTENTS

There are eleven groups of problems in this module. Each group consists of five questions in multiple-choice format. The groups are labeled A through K. A number of the problems are considerably longer and more tedious than what would be encountered on the actual exam. However, they cover necessary subject areas and will provide you with important background information.

## A.

Two different types of structures are shown in elevation. For both structures, the heights and floor weights are the same. Each building will be occupied by 6000 or more people. Use seismic zone 3 for building I and seismic zone 4 for building II. The value of $N_a$ and $N_v$ (near-source factors) are 1.0 for buildings I and II. The penthouse is a special moment-resisting steel frame.

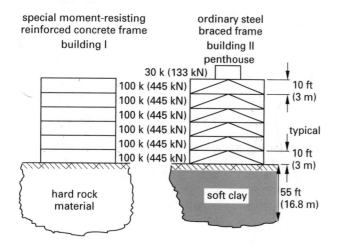

1. For building I, determine the base shear consistent with the seismic analysis required in UBC Sec. 1630.

    A.  16 k  (71 kN)
    B.  26 k  (118 kN)
    C.  42 k  (188 kN)
    D.  50 k  (220 kN)

2. For building II, determine the base shear consistent with the seismic analysis required in UBC Sec. 1630.

    A.  25 k  (111 kN)
    B.  100 k  (450 kN)
    C.  140 k  (626 kN)
    D.  250 k  (1118 kN)

3. The fundamental period of building I is 0.6 sec. Based on UBC requirements, what should the allowable inter-story displacement (drift) be?

    A.  0.30 in  (8 mm)
    B.  0.4 in  (10 mm)
    C.  0.5 in  (13 mm)
    D.  0.60 in  (15 mm)

4. Assume the typical height of each floor is 15 ft (4.6 m) instead of 10 ft (3 m). For building I, determine the force at roof level, $F_t$.

    A.  0 lbf  (0 N)
    B.  1190 lbf  (5280 N)
    C.  1490 lbf  (6630 N)
    D.  4810 lbf  (21 425 N)

5. Determine the seismic loading, $F_p$, for the penthouse. The penthouse is rigidly attached to the main building.

    A.  10,800 lbf  (47 880 N)
    B.  22,100 lbf  (98 000 N)
    C.  43,200 lbf  (191 500 N)
    D.  68,400 lbf  (303 200 N)

## B.

For an office building in seismic zone 4, an architectural firm proposes the ordinary steel braced frame structure shown. The weights and heights of each story are as illustrated. The proximity of the building to the seismic source type $A$ is estimated to be 6.2 mi (10 km). Ignore vertical structural irregularities for this building. Use the static lateral-force procedure.

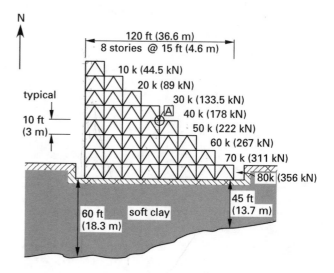

6. Based on UBC seismic design requirements, what is the base shear for this building?

    A.  14,250 lbf  (63 400 N)
    B.  24,700 lbf  (109 800 N)
    C.  41,300 lbf  (183 700 N)
    D.  57,850 lbf  (257 300 N)

7. An architect wants to increase the story heights for all the floors without creating any additional force ($F_t$) at the roof level. For this modification, what should the maximum height of the structure be?

    A.  70 ft  (21.3 m)
    B.  80 ft  (24.4 m)
    C.  107 ft  (32.7 m)
    D.  114 ft  (34.8 m)

8. At point A as marked in the illustration, determine the magnitude of the overturning moment. Assume the calculated base shear is 49.5 k (220 kN).

    A.  2000 ft-lbf  (2710 N·m)
    B.  31,000 ft-lbf  (42 030 N·m)
    C.  268,000 ft-lbf  (363 360 N·m)
    D.  536,000 ft-lbf  (723 700 N·m)

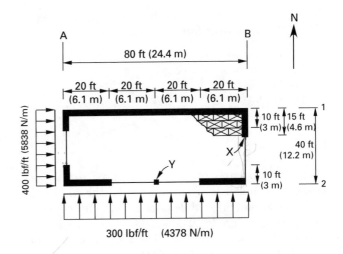

9. Assume the calculated base shear is 49.5 k (220 kN). Determine the total overturning moment at the base.

    A.  $1.0 \times 10^3$ ft-k  ($1.4 \times 10^3$ kN·m)
    B.  $1.5 \times 10^3$ ft-k  ($2.0 \times 10^3$ kN·m)
    C.  $2.2 \times 10^3$ ft-k  ($3.0 \times 10^3$ kN·m)
    D.  $3.0 \times 10^3$ ft-k  ($4.1 \times 10^3$ kN·m)

10. Determine the design uplift against the overturning moment. Assume the calculated base shear is 49.5 k (220 kN).

    A.  8000 ft-k  (10 850 kN·m)
    B.  13,800 ft-k  (18 670 kN·m)
    C.  21,000 ft-k  (28 470 kN·m)
    D.  27,900 ft-k  (37 830 kN·m)

## C.

A one-story, wood-frame restaurant is shown in plan view. The height of the wood structural panel shear walls is 16 ft (4.9 m). The roof dead load and wall dead load are 24 lbf/ft² (1149 N/m²) and 20 lbf/ft² (958 N/m²), respectively. The seismic loadings to the diaphragm are as given in the following illustration. ($\rho = 1.0$)

11. Along line A (the west wall), determine the strut load for the shear walls.

    A.  0 lbf  (0 N)
    B.  3000 lbf  (13 340 N)
    C.  6000 lbf  (26 690 N)
    D.  12,000 lbf  (53 380 N)

12. For the 25 ft (7.6 m) opening at line B, determine the magnitude of the drag strut force that will be transmitted to the shear wall panels.

    A.  0 lbf  (0 N)
    B.  1500 lbf  (6670 N)
    C.  3750 lbf  (16 680 N)
    D.  7500 lbf  (33 360 N)

13. Determine the maximum drag strut force at point Y along line 2.

    A.  0 lbf  (0 N)
    B.  2000 lbf  (8900 N)
    C.  4000 lbf  (17 790 N)
    D.  8000 lbf  (35 590 N)

14. At point X, what is the maximum drag strut force?

    A.  3750 lbf  (16 680 N)
    B.  7500 lbf  (33 360 N)
    C.  12,000 lbf  (53 380 N)
    D.  24,000 lbf  (106 760 N)

15. When the north-south loading doubles, the magnitude of the drag strut force at point Y

    A.　halves.
    B.　experiences no change.
    C.　doubles.
    D.　triples.

# D.

There are two water tanks and one guyed tower stack as shown. Tank I is well anchored on the top of the steel braced building, while tank II is rigidly bolted to a foundation at grade. The weight of the guyed tower stack is 15,000 lbf (66 723 N). The weights of the tanks when full are 30,000 lbf (133 446 N) and 100,000 lbf (444 820 N), respectively. They all are located in seismic zone 4. Assume sufficient freeboard exists for both tanks. The proximity of these structures to the seismic source type B is estimated to be 6.2 mi (10 km).

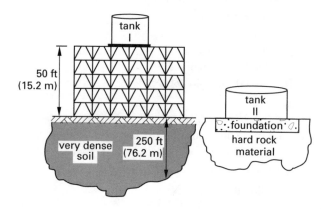

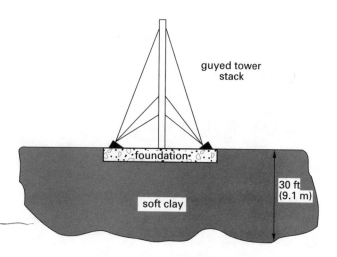

16. Assume the fundamental period of tank I is 0.04 sec (0.04 s). Based on UBC Chapter 16 requirements, determine the lateral force on tank I.

    A.　6720 lbf　(29 890 N)
    B.　8400 lbf　(37 360 N)
    C.　48,000 lbf　(213 510 N)
    D.　190,910 lbf　(849 190 N)

17. Assume the fundamental period of tank II is 0.05 sec (0.05 s). This tank is a nonbuilding structure. Based on UBC Chapter 16 requirements, determine the lateral force on tank II.

    A.　17,900 lbf　(79 600 N)
    B.　22,400 lbf　(99 600 N)
    C.　128,000 lbf　(569 400 N)
    D.　290,910 lbf　(1 071 700 N)

18. Assume the fundamental period of guyed tower stack is 0.01 sec (0.01 s). Based on UBC Chapter 16 requirements, what should the design lateral force be?

    A.　3020 lbf　(13 430 N)
    B.　3780 lbf　(16 810 N)
    C.　21,600 lbf　(9680 N)
    D.　49,660 lbf　(220 900 N)

19. Assume guyed tower stack is a nonbuilding structure with a fundamental period of 0.12 sec (0.12 s). Based on UBC requirements, what should the design lateral force be?

    A.　3000 lbf　(13 450 N)
    B.　3300 lbf　(14 720 N)
    C.　4650 lbf　(20 700 N)
    D.　41,400 lbf　(184 060 N)

20. Assume the contents of tank II are highly toxic, and would be hazardous to the safety of the general public if released. This tank is a nonbuilding structure. The fundamental period is 0.04 sec. Based on UBC requirements, what should the design lateral force be?

    A.　22,400 lbf　(99 640 N)
    B.　28,000 lbf　(124 550 N)
    C.　128,000 lbf　(569 370 N)
    D.　192,000 lbf　(854 050 N)

# E.

As shown in the following illustration, the reinforced masonry wall of a storage building spans vertically between the floor and the roof. The masonry-walled building with a wood structural panel (flexible) roof diaphragm is located in seismic zone 4. The wall weighs 100 lbf/ft$^2$ (4788 N/m$^2$) and the roof weighs 20 lbf/ft$^2$ (958 N/m$^2$). The reinforcing steel is ASTM A615, grade 60. The earthquake loading is in the north-south direction. Use simplified design base shear procedure. Assume the seismic source type is A and the proximity of the building to the seismic source is 3.1 mi (5 km). Disregard all live loads.

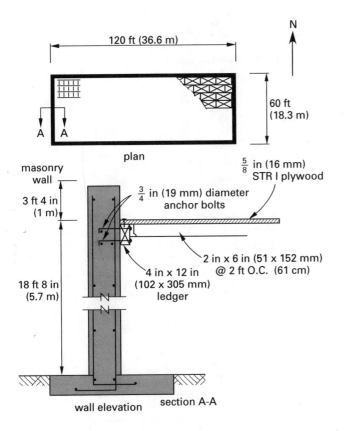

120 ft (36.6 m)

60 ft (18.3 m)

A A

plan

$\frac{5}{8}$ in (16 mm) STR I plywood

masonry wall

3 ft 4 in (1 m)

$\frac{3}{4}$ in (19 mm) diameter anchor bolts

2 in x 6 in (51 x 152 mm) @ 2 ft O.C. (61 cm)

4 in x 12 in (102 x 305 mm) ledger

18 ft 8 in (5.7 m)

wall elevation    section A-A

21. The wall and its connections to the roof should be designed to resist the lateral force normal to the wall face, calculated by the appropriate UBC formula. Choose the correct formula.

A. $0.5C_aIWp_x$
B. $1.0C_aIWp_x$
C. $4.0C_aIpWp$
D. $C_vIW/RT$

22. For the north-south direction, determine the seismic loading to the roof diaphragm.

A. 920 lbf/ft   (13 430 N/m)
B. 1320 lbf/ft   (19 260 N/m)
C. 2640 lbf/ft   (38 530 N/m)
D. 3540 lbf/ft   (51 660 N/m)

23. Assuming that anchorage points are 4 ft (1.2 m) O.C., determine the required anchorage force at the roof line. Use UBC Formula 32-2.

A. 1120 lbf   (4860 N)
B. 1730 lbf   (7510 N)
C. 2350 lbf   (10 200 N)
D. 5200 lbf   (22 570 N)

24. For a seismic loading perpendicular to a wall, determine the required anchorage force at the bottom of the wall. Use UBC Formula 32-2.

A. 140 lbf/ft   (2040 N/m)
B. 280 lbf/ft   (4070 N/m)
C. 560 lbf/ft   (8150 N/m)
D. 750 lbf/ft   (10 910 N/m)

25. For north-south earthquake forces, determine the minimum required chord reinforcement area, $A_s$.

A. 0.6 in$^2$   (3.9 cm$^2$)
B. 1.0 in$^2$   (6.5 cm$^2$)
C. 1.2 in$^2$   (8.0 cm$^2$)
D. 1.6 in$^2$   (10.3 cm$^2$)

# F.

The south wall of a one-story commercial building in San Francisco is shown as follows in elevation with the given loading. The building has a rigid roof diaphragm that is adequately anchored to the walls. The plywood shear wall panels have a uniform thickness and the steel frames have the same properties.

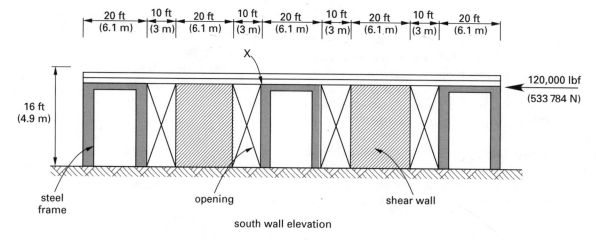

south wall elevation

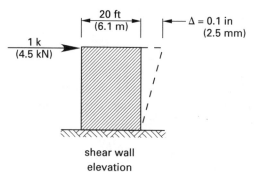

shear wall
elevation

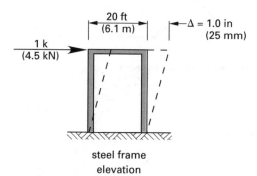

steel frame
elevation

26. Determine the relative rigidity of the south wall.

A. 5
B. 15
C. 23
D. 32

27. Each of the wood structural shear wall panels carries a load equal to

A. 24,000 lbf   (107 000 N).
B. 30,000 lbf   (133 850 N).
C. 48,000 lbf   (214 150 N).
D. 52,000 lbf   (232 000 N).

28. Each of the steel frames carries a load equal to

A. 5200 lbf   (23 210 N).
B. 15,600 lbf   (69 630 N).
C. 30,000 lbf   (133 910 N).
D. 48,000 lbf   (214 250 N).

29. Assume that a concrete shear wall with a relative rigidity of 5 replaces the existing center steel frame. What load is carried by the concrete shear wall?

A. 22,000 lbf   (97 860 N)
B. 24,000 lbf   (106 760 N)
C. 30,000 lbf   (133 450 N)
D. 40,000 lbf   (177 930 N)

30. Assume a wood structural panel roof diaphragm for this building. Determine the magnitude of the drag strut force at point X.

A. 3100 lbf   (13 400 N)
B. 6200 lbf   (26 800 N)
C. 9100 lbf   (39 340 N)
D. 18,200 lbf   (78 670 N)

## G.

The plan view of a one-story office building underlain with type $S_A$ soil profile in seismic zone 3 is shown. The building has a wood structural panel roof diaphragm and shear walls. The roof drag struts are continuous. The roof dead load is 20 lbf/ft$^2$ (958 N/m$^2$) and the wall dead load is 16 lbf/ft$^2$ (766 N/m$^2$). The building is 14 ft (4.3 m) tall. Use the UBC simplified design base shear procedure.

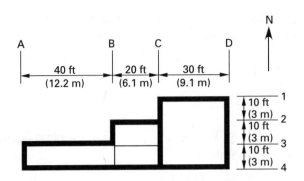

# H.

A proposed one-story motel with the wood structural panel roof diaphragm and walls is shown in plan view. The roof diaphragm is adequately anchored to the shear walls. The north-south and east-west lateral loadings to the diaphragm are 400 lbf/ft (5838 N/m) and 160 lbf/ft (2335 N/m), respectively. ($\rho = 1.0$)

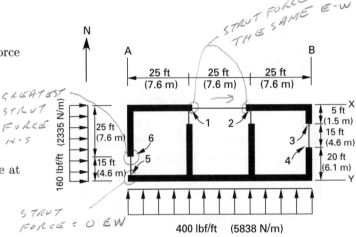

31. For north-south loading, what is the roof shear force at line B?

    A. 1200 lbf   (5350 N)
    B. 1600 lbf   (7130 N)
    C. 1750 lbf   (7800 N)
    D. 1910 lbf   (8515 N)

32. For north-south loading, what is the shear force at the mid-height of the shear wall at line C?

    A. 85 lbf/ft   (1240 N/m)
    B. 95 lbf/ft   (1390 N/m)
    C. 115 lbf/ft   (1670 N/m)
    D. 130 lbf/ft   (1890 N/m)

33. At line 4 between lines A and B, determine the maximum magnitude of the chord force.

    A. 215 lbf   (970 N)
    B. 425 lbf   (1920 N)
    C. 1100 lbf   (4970 N)
    D. 1750 lbf   (7910 N)

34. Determine the maximum magnitude of the chord force at line 4 between lines B and C.

    A. 200 lbf   (910 N)
    B. 425 lbf   (1930 N)
    C. 1200 lbf   (5460 N)
    D. 1750 lbf   (7960 N)

35. Determine the maximum magnitude of the chord force at line 1 between lines D and C.

    A. 215 lbf   (960 N)
    B. 400 lbf   (1780 N)
    C. 1200 lbf   (5340 N)
    D. 1750 lbf   (7780 N)

36. For the east-west loading direction, determine the magnitude of the strut force at point 1.

    A. 325 lbf   (1470 N)
    B. 450 lbf   (2030 N)
    C. 535 lbf   (2370 N)
    D. 1050 lbf   (4740 N)

37. For the east-west loading direction, what is the magnitude of the strut force along line Y?

    A. 0 lbf   (0 N)
    B. 2000 lbf   (8900 N)
    C. 3200 lbf   (14 230 N)
    D. 6400 lbf   (28 470 N)

38. For the east-west loading direction, what is the magnitude of the strut force at point 2?

    A. 175 lbf   (760 N)
    B. 350 lbf   (1510 N)
    C. 535 lbf   (2370 N)
    D. 1100 lbf   (4760 N)

39. For the north-south loading direction, at what location is the magnitude of the strut force greatest?

    A. at point 3
    B. at point 4
    C. at point 5
    D. at point 6

40. For east-west loading, how much of the horizontal diaphragm shear force should be transferred to both shear walls at line X by the drag strut between points 1 and 2?

    A. 525 lbf (2320 N)
    B. 550 lbf (2430 N)
    C. 1070 lbf (4750 N)
    D. 3200 lbf (14 140 N)

## I.

A one-story commercial wood-frame building is shown in the following illustration. The height is 12 ft (3.7 m). The dead load of the walls is 16 lbf/ft$^2$ (766 N/m$^2$). ($\rho = 1.0$)

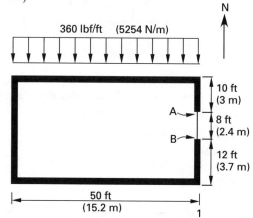

41. Determine the magnitude of the drag strut force over the 8 ft (2.4 m) opening at line 1.

    A. 0 lbf (0 N)
    B. 1125 lbf (4930 N)
    C. 2400 lbf (10 530 N)
    D. 3270 lbf (14 340 N)

42. Determine the magnitude of the drag strut force at point A on line 1.

    A. 750 lbf (3240 N)
    B. 900 lbf (3890 N)
    C. 1100 lbf (4750 N)
    D. 1500 lbf (6480 N)

43. For the east shear wall, the 10 ft (3 m) shear panel resists what proportion of the applied diaphragm load?

    A. 33%
    B. 40%
    C. 45%
    D. 50%

44. For the 12 ft (3.7 m) shear panel, determine the overturning moment.

    A. 43 ft-k (60 kN·m)
    B. 49 ft-k (68 kN·m)
    C. 59 ft-k (82 kN·m)
    D. 64 ft-k (89 kN·m)

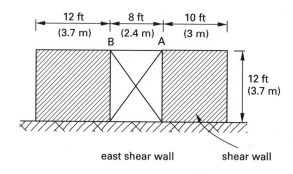

east shear wall      shear wall

45. Assume there is no opening on the east shear wall. The entire wall has a uniform thickness. Ignore any roof dead load contribution to the wall's resistance to overturning. Based on UBC requirements, compute the resisting moment.

    A. 37 ft-k (50 kN·m)
    B. 43 ft-k (58 kN·m)
    C. 78 ft-k (106 kN·m)
    D. 86 ft-k (117 kN·m)

## J.

A one-story wood-frame building is shown in the following illustration. The height of the shear walls is 14 ft (4.3 m) and the walls are uniform throughout their entire length. The wall dead load is 16 lbf/ft$^2$ (766 N/m$^2$) and the seismic base shear coefficient $3.0C_a/R$ for the UBC simplified design base shear procedure is 0.1375. ($\rho = 1.0$)

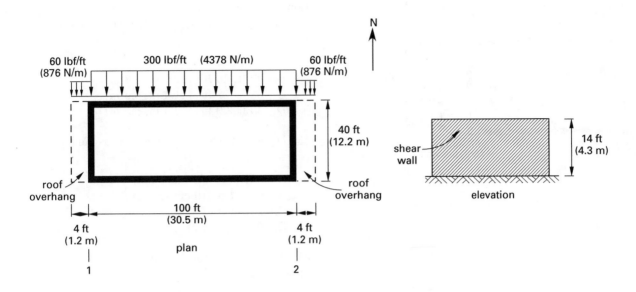

46. For a seismic loading in the north-south direction, what is the diaphragm shear force at line 1?

    A.  15,000 lbf  (66 800 N)
    B.  15,120 lbf  (67 280 N)
    C.  15,240 lbf  (67 815 N)
    D.  15,360 lbf  (68 350 N)

47. For a north-south direction, what is the unit diaphragm shear stress at line 2?

    A.  378 lbf/ft  (5520 N/m)
    B.  381 lbf/ft  (5560 N/m)
    C.  384 lbf/ft  (5600 N/m)
    D.  387 lbf/ft  (5650 N/m)

48. Determine the shear stress in the shear wall panel along line 1.

    A.  360 lbf/ft  (5250 N/m)
    B.  381 lbf/ft  (5560 N/m)
    C.  400 lbf/ft  (5840 N/m)
    D.  415 lbf/ft  (6060 N/m)

49. Using UBC Table 23-II-H, assume Case 2 for the roof wood structural panel layout, 15/32 in (12 mm) (structural I) for the roof sheathing, 10d common nails, and 2 in (5 cm) nominal framing. Blocking is required. At diaphragm boundaries along line 1, what should the nail spacing be?

    A.  2.0 in  (5 cm)
    B.  2.5 in  (6 cm)
    C.  4.0 in  (10 cm)
    D.  6.0 in  (15 cm)

50. For the shear panels, $\frac{3}{8}$ in (1 cm) panel grade C-D sheathing and 8d common nails are used. What should the nail spacing be at the plywood panel edges for the wall along line 2? The studs are spaced at 24 in (0.61 m) on center.

    A.  2.0 in  (5 cm)
    B.  3.0 in  (8 cm)
    C.  4.0 in  (10 cm)
    D.  6.0 in  (15 cm)

## K.

A single-story building with concrete shear walls and a roof diaphragm is shown. The story height is 12 ft (3.7 m). The thickness of the roof and walls are uniform. The roof and wall dead loads are 25 lbf/ft$^2$ (1197 N/m$^2$) and 50 lbf/ft$^2$ (2394 N/m$^2$), respectively.

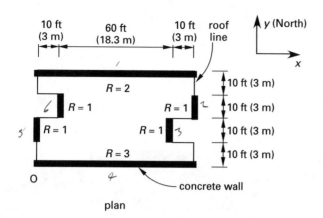

plan

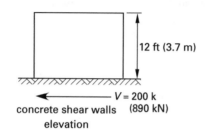

12 ft (3.7 m)

V = 200 k
concrete shear walls     (890 kN)
elevation

51. Determine the location of the center of mass, denoted by $\bar{x}$ and $\bar{y}$, respectively. Assume that point O is the origin. Disregard accidental torsion.

   A. 0 ft, 0 ft  (0 m, 0 m)
   B. 38 ft, 19 ft  (11.4 m, 5.7 m)
   C. 42 ft, 23 ft  (12.6 m, 6.9 m)
   D. 40 ft, 20 ft  (12.0 m, 6.0 m)

52. The rigidities of the walls are as given in the illustration. Determine the location of the center of rigidity, denoted by $(\bar{x}_R$ and $\bar{y}_R)$. Use the lower left corner, point O, as the origin.

   A. 0 ft, 0 ft  (0 m, 0 m)
   B. 40 ft, 16 ft  (12.0 m, 4.8 m)
   C. 38 ft, 22 ft  (11.4 m, 6.6 m)
   D. 40 ft, 20 ft  (12.0 m, 6.0 m)

53. Determine the eccentricity in $x$- and $y$-direction, respectively.

   A. 0 ft, 0 ft  (0 cm, 0 cm)
   B. 0 ft, 4 ft  (0 cm, 120 cm)
   C. 1 ft, 1 ft  (30 cm, 30 cm)
   D. 2 ft, 2 ft  (60 cm, 60 cm)

54. For the north-south direction, what will the design torsional moment be?

   A. 600 ft-k  (810 kN·m)
   B. 800 ft-k  (1070 kN·m)
   C. 1000 ft-k  (1340 kN·m)
   D. 1200 ft-k  (1610 kN·m)

55. Determine the design torsional moment in the east-west direction.

   A. 800 ft-k  (1060 kN·m)
   B. 1000 ft-k  (1330 kN·m)
   C. 1200 ft-k  (1600 kN·m)
   D. 1400 ft-k  (1860 kN·m)

In solutions C (11–15), H (36–40), I (41–45), and J (46–50), $\rho$ represents a reliability/redundancy factor that should be assigned to all structures according to the UBC [Sec. 1630.1.1]. This factor is based on the extent of structural redundancy inherent in the design configuration of the structure and its lateral force-resisting system.

## A.

1. Answer B

*SI solution*

From UBC Table 16-I, $Z = 0.3$. From UBC Table 16-K, $I = 1.0$. From UBC Table 16-J, the soil profile type is $S_A$. From UBC Table 16-N, $R = 8.5$. From UBC Tables 16-Q and 16-R, the seismic response coefficients $C_a$ and $C_v$ are 0.24. From the UBC [Sec. 1630.2.2, Item 1], $C_t = 0.0731$.

$$h_n = (6 \text{ stories})(3 \text{ m})$$
$$= 18 \text{ m}$$
$$T = C_t(h_n)^{3/4}$$
$$= (0.0731)(18 \text{ m})^{3/4}$$
$$= 0.64 \text{ s}$$

From the UBC [Sec. 1630.5], since 0.64 s < 0.7 s, $F_t = 0$.

$$W = (6 \text{ stories})(445 \text{ kN})$$
$$= 2670 \text{ kN}$$

Use UBC Formula 30-4.

$$V = \left(\frac{C_v I}{RT}\right) W$$
$$= \left(\frac{(0.24)(1.0)}{(8.5)(0.64)}\right)(2670 \text{ kN})$$
$$= 117.8 \text{ kN}$$

Based on the UBC [Sec. 1630.2.1], the total design base shear should not exceed $2.5 C_a IW/R$ and should not be less than $0.11 C_a IW$.

$$V_{\max} = \frac{(2.5)(0.24)(1.0)(2670 \text{ kN})}{8.5}$$
$$= 188.5 \text{ kN}$$
$$V_{\min} = (0.11)(0.24)(1.0)(2670 \text{ kN})$$
$$= 70.5 \text{ kN}$$

Since 70.5 kN < 117.8 kN < 188.5 kN, the value of $V = 117.8$ kN may be used.

*Customary U.S. solution*

From UBC Table 16-I, $Z = 0.3$. From UBC Table 16-K, $I = 1.0$. From UBC Table 16-J, the soil profile type is $S_A$. From UBC Table 16-N, $R = 8.5$. From UBC Tables 16-Q and 16-R, the seismic response coefficients $C_a$ and $C_v$ are 0.24. From the UBC [Sec. 1630.2.2, Item 1], $C_t = 0.030$.

$$h_n = (6 \text{ stories})(10 \text{ ft})$$
$$= 60 \text{ ft}$$
$$T = C_t(h_n)^{3/4}$$
$$= (0.030)(60 \text{ ft})^{3/4}$$
$$= 0.65 \text{ sec}$$

From the UBC [Sec. 1630.5], since 0.65 sec < 0.7 sec, $F_t = 0$.

$$W = (6 \text{ stories})(100 \text{ k})$$
$$= 600 \text{ k}$$

Use UBC Formula 30-4.

$$V = \left(\frac{C_v I}{RT}\right) W$$

$$= \left(\frac{(0.24)(1.0)}{(8.5)(0.65)}\right)(600 \text{ k})$$

$$= 26.06 \text{ k}$$

Based on the UBC [Sec. 1630.2.1], the total design base shear should not exceed $2.5C_a IW/R$ and should not be less than $0.11C_a IW$.

$$V_{\max} = \frac{(2.5)(0.24)(1.0)(600 \text{ k})}{8.5}$$

$$= 42.35 \text{ k}$$

$$V_{\min} = (0.11)(0.24)(1.0)(600 \text{ k})$$

$$= 15.84 \text{ k}$$

Since 15.84 k $<$ 26.06 k $<$ 42.35 k, the value of $V = 26.06$ k may be used.

2. Answer B

*SI solution*

From UBC Table 16-I, $Z$ is 0.4. From UBC Table 16-K, $I$ is 1.0. From UBC Table 16-J, the soil profile type is $S_E$. From UBC Table 16-N, $R$ is 5.6. From UBC Table 16-Q, the seismic response coefficient $C_a = 0.36N_a$. From UBC Table 16-R, the seismic response coefficient $C_v = 0.96N_v$. From the UBC [Sec. 1630.2.2, Item 1], $C_t$ is 0.0488.

$$h_n = (6 \text{ stories})(3 \text{ m})$$

$$= 18 \text{ m}$$

$$T = C_t(h_n)^{3/4}$$

$$= (0.0488)(18 \text{ m})^{3/4}$$

$$= 0.43 \text{ s}$$

From the UBC [Sec. 1630.5], since 0.43 s $<$ 0.7 s, $F_t = 0$.

$$W = (6 \text{ stories})(445 \text{ kN}) + (133 \text{ kN})$$

$$= 2803 \text{ kN}$$

Use UBC Formula 30-4.

$$V = \left(\frac{C_v I}{RT}\right) W$$

$$C_v = 0.96N_v$$

$$= (0.96)(1.0)$$

$$= 0.96$$

$$V = \left(\frac{(0.96)(1.0)}{(5.6)(0.43)}\right)(2803 \text{ kN})$$

$$= 1117.5 \text{ kN}$$

Based on the UBC [Sec. 1630.2.1], the total design base shear need not exceed $2.5C_a IW/R$ and should not be less than $0.11C_a IW$.

$$C_a = 0.36N_a$$

$$= (0.36)(1.0)$$

$$= 0.36$$

$$V_{\max} = \frac{(2.5)(0.36)(1.0)(2803 \text{ kN})}{5.6}$$

$$= 450.5 \text{ kN}$$

$$V_{\min} = (0.11)(0.36)(1.0)(2803 \text{ kN})$$

$$= 111.0 \text{ kN}$$

The UBC [Sec. 1630.2.1] also requires that the total base shear for seismic zone 4 not be less than $0.8ZN_v IW/R$.

For seismic zone 4,

$$V_{\min} = \frac{(0.8)(0.4)(1.0)(2803 \text{ kN})}{5.6}$$

$$= 160.2 \text{ kN}$$

Since 1117.5 kN $>$ 450.5 kN ($V_{\max}$), the value of $V = 450.5$ kN can be used.

*Customary U.S. solution*

From UBC Table 16-I, $Z$ is 0.4. From UBC Table 16-K, $I$ is 1.0. From UBC Table 16-J, the soil profile type is $S_E$. From UBC Table 16-N, $R$ is 5.6. From UBC Table 16-Q, the seismic response coefficient $C_a = 0.36N_a$. From UBC Table 16-R, the seismic response coefficient $C_v = 0.96N_v$. From the UBC [Sec. 1630.2.2, Item 1], $C_t$ is 0.020.

$$h_n = (6 \text{ stories})(10 \text{ ft})$$

$$= 60 \text{ ft}$$

$$T = C_t(h_n)^{3/4}$$

$$= (0.020)(60 \text{ ft})^{3/4}$$

$$= 0.43 \text{ sec}$$

From the UBC [Sec. 1630.5], since 0.43 sec $<$ 0.7 sec, $F_t = 0$.

$$W = (6 \text{ stories})(100 \text{ k}) + (30 \text{ k})$$

$$= 630 \text{ k}$$

Use UBC Formula 30-4.

$$V = \left(\frac{C_v I}{RT}\right) W$$

$$C_v = 0.96 N_v$$
$$= (0.96)(1.0)$$
$$= 0.96$$

$$V = \left(\frac{(0.96)(1.0)}{(5.6)(0.43)}\right)(630 \text{ k})$$
$$= 251.2 \text{ k}$$

Based on the UBC [Sec. 1630.2.1], the total design base shear need not exceed $2.5C_a IW/R$ and should not be less than $0.11C_a IW$.

$$C_a = 0.36 N_a$$
$$= (0.36)(1.0)$$
$$= 0.36$$

$$V_{\max} = \frac{(2.5)(0.36)(1.0)(630 \text{ k})}{5.6}$$
$$= 101.3 \text{ k}$$

$$V_{\min} = (0.11)(0.36)(1.0)(630 \text{ k})$$
$$= 24.9 \text{ k}$$

The UBC [Sec. 1630.2.1] also requires that the total base shear for seismic zone 4 should not be less than $0.8ZN_v IW/R$.

For seismic zone 4,

$$V_{\min} = \frac{(0.8)(0.4)(1.0)(630 \text{ k})}{5.6}$$
$$= 36.0 \text{ k}$$

Since $251.2 \text{ k} > 101.3 \text{ k}$ $(V_{\max})$, the value of $V = 101.3 \text{ k}$ can be used.

**3.** Answer C

*SI solution*

$$h_x = 3 \text{ m}$$
$$T = 0.6 \text{ s}$$

Based on the UBC [Sec. 1630.10.2], the story drift using $\Delta_M$ for structures having a building period of less than 0.7 s should be limited to

$$\Delta_M = 0.025 h_x$$
$$= (0.025)(3 \text{ m})\left(1000 \frac{\text{mm}}{\text{m}}\right)$$
$$= 75 \text{ mm}$$

For the allowable interstory displacement (drift), use UBC Formula 30-17.

$$\Delta_M = 0.7R\Delta_S$$

From UBC Table 16-N, $R = 8.5$.

$$\Delta_S = \frac{75 \text{ mm}}{(0.7)(8.5)} = 12.6 \text{ mm}$$
$$\approx 13 \text{ mm}$$

*Customary U.S. solution*

$$h_x = 10 \text{ ft}$$
$$T = 0.6 \text{ sec}$$

Based on the UBC [Sec. 1630.10.2], story drift using $\Delta_M$ for structures having a building period of less than 0.7 sec should be limited to

$$\Delta_M = 0.025 h_x$$
$$= (0.025)(10 \text{ ft})\left(12 \frac{\text{in}}{\text{ft}}\right)$$
$$= 3.0 \text{ in}$$

For the allowable interstory displacement (drift), use UBC Formula 30-17.

$$\Delta_M = 0.7R\Delta_S$$

From UBC Table 16-N, $R = 8.5$.

$$\Delta_S = \frac{3.0 \text{ in}}{(0.7)(8.5)} = 0.5 \text{ in}$$

**4.** Answer B

*SI solution*

Use UBC Formula 30-14.

$$F_t = 0.07TV$$

For building I,

$$h_n = (6 \text{ stories})(4.6 \text{ m})$$
$$= 27.6 \text{ m}$$
$$T = C_t(h_n)^{3/4}$$
$$= (0.0731)(27.6 \text{ m})^{3/4}$$
$$= 0.88 \text{ s}$$

Since 0.88 s > 0.7 s, $F_t \neq 0$.

$$W = (6 \text{ stories})(445 \text{ kN})$$
$$= 2670 \text{ kN}$$

Use UBC Formula 30-4.

$$V = \left(\frac{C_v I}{RT}\right) W$$

For building I, $C_v = 0.24$, $I = 1.0$, $R = 8.5$, and $T = 0.88$ s.

$$V = \left(\frac{(0.24)(1.0)}{(8.5)(0.88)}\right)(2670 \text{ kN})$$
$$= 85.7 \text{ kN}$$
$$F_t = 0.07TV$$
$$= (0.07)(0.88)(85.7 \text{ kN})\left(1000 \frac{\text{N}}{\text{kN}}\right)$$
$$= 5279 \text{ N}$$

From the UBC [Sec. 1630.5], $F_t$ need not exceed 0.25V.

$$0.25V = (0.25)(85.7 \text{ kN})\left(1000 \frac{\text{N}}{\text{kN}}\right)$$
$$= 21\,425 \text{ N}$$

Since 5279 N < 21 425 N, the value of $F_t = 5279$ N should be used.

*Customary U.S. solution*

Use UBC Formula 30-14.

$$F_t = 0.07TV$$

For building I,

$$h_n = (6 \text{ stories})(15 \text{ ft})$$
$$= 90 \text{ ft}$$
$$T = C_t(h_n)^{3/4}$$
$$= (0.030)(90 \text{ ft})^{3/4}$$
$$= 0.88 \text{ sec}$$

Since 0.88 sec > 0.7 sec, $F_t \neq 0$.

$$W = (6 \text{ stories})(100 \text{ k})$$
$$= 600 \text{ k}$$

Use UBC Formula 30-4.

$$V = \left(\frac{C_v I}{RT}\right) W$$

For building I, $C_v = 0.24$, $I = 1.0$, $R = 8.5$, and $T = 0.88$ sec.

$$V = \left(\frac{(0.24)(1.0)}{(8.5)(0.88)}\right)(600 \text{ k})$$
$$= 19.25 \text{ k}$$
$$F_t = 0.07TV$$
$$= (0.07)(0.88)(19.25 \text{ k})\left(1000 \frac{\text{lbf}}{\text{k}}\right)$$
$$= 1186 \text{ lbf}$$

From the UBC [Sec. 1630.5], $F_t$ need not exceed 0.25V.

$$0.25V = (0.25)(19.25 \text{ k})\left(1000 \frac{\text{lbf}}{\text{k}}\right)$$
$$= 4813 \text{ lbf}$$

Since 1186 lbf < 4813 lbf, the value of $F_t = 1186$ lbf should be used.

5. Answer C

*SI solution*

Use UBC Formula 32-1.

$$F_p = 4.0 C_a I_p W_p$$

From UBC Table 16-I, $Z$ is 0.4. From UBC Table 16-K, $I_p$ is 1.0. From UBC Table 16-J, the soil profile type is $S_E$. From UBC Table 16-Q, the seismic response coefficient $C_a$ is

$$C_a = 0.36 N_a$$
$$= (0.36)(1.0)$$
$$= 0.36$$
$$F_p = (4.0)(0.36)(1.0)(133 \text{ kN})$$
$$= 191.5 \text{ kN}$$
$$= (191.5 \text{ kN})\left(1000 \frac{\text{N}}{\text{kN}}\right)$$
$$= 191\,500 \text{ N}$$

*Customary U.S. solution*

Use UBC Formula 32-1.

$$F_p = 4.0C_aI_pW_p$$

From UBC Table 16-I, $Z$ is 0.4. From UBC Table 16-K, $I_p$ is 1.0. From UBC Table 16-J, the soil profile type is $S_E$. From UBC Table 16-Q, the seismic response coefficient $C_a$ is

$$C_a = 0.36N_a$$
$$= (0.36)(1.0)$$
$$= 0.36$$
$$F_p = (4.0)(0.36)(1.0)(30\text{ k})$$
$$= 43.2\text{ k}$$
$$= (43.2\text{ k})\left(1000\ \frac{\text{lbf}}{\text{k}}\right)$$
$$= 43{,}200\text{ lbf}$$

# B.

6. Answer D

*SI solution*

From UBC Table 16-I, $Z$ is 0.4. From UBC Table 16-K, $I$ is 1.0. From UBC Table 16-J, the soil profile type is $S_E$. From UBC Table 16-N, $R$ is 5.6. From UBC Table 16-Q, $C_a$ is $0.36N_a$. From UBC Table 16-R, $C_v$ is 0.96 $N_v$. From UBC Table 16-S, the near-source factor $N_a$ is 1.0. From UBC Table 16-T, the near-source factor $N_v$ is 1.2. From the UBC [Sec. 1630.2.2, Item 1], $C_t$ is 0.0488.

$$h_n = (8\text{ stories})(3\text{ m}) = 24\text{ m}$$
$$T = C_t(h_n)^{3/4}$$
$$= (0.0488)(24\text{ m})^{3/4}$$
$$= 0.53\text{ s}$$

Since 0.53 s < 0.7 s, $F_t = 0$.

$$W = 44.5\text{ kN} + 89.0\text{ kN} + 133.5\text{ kN}$$
$$+ 178.0\text{ kN} + 222.0\text{ kN} + 267.0\text{ kN}$$
$$+ 311.0\text{ kN} + 356.0\text{ kN}$$
$$= 1601\text{ kN}$$
$$C_a = 0.36N_a$$
$$= (0.36)(1.0)$$
$$= 0.36$$
$$C_v = 0.96N_v$$
$$= (0.96)(1.2)$$
$$= 1.15$$

Use UBC Formula 30-4.

$$V = \left(\frac{C_vI}{RT}\right)W$$
$$= \left(\frac{(1.15)(1.0)}{(5.6)(0.53)}\right)(1601\text{ kN})$$
$$= 620.3\text{ kN}$$
$$= 620\,300\text{ N}$$

Based on the UBC [Sec. 1630.2.1], the total design base shear need not exceed $2.5C_aIW/R$ and should not be less than $0.11C_aIW$.

$$V_{\max} = \frac{(2.5)(0.36)(1.0)(1601\text{ kN})}{5.6}$$
$$= 257.3\text{ kN}$$
$$= 257\,300\text{ N}$$
$$V_{\min} = (0.11)(0.36)(1.0)(1601\text{ kN})\left(1000\ \frac{\text{N}}{\text{kN}}\right)$$
$$= 63\,400\text{ N}$$

Specifically for seismic zone 4, the UBC [Sec. 1630.2.1] requires that total base shear not be less than $0.8ZN_vIW/R$.

$$V_{\min(\text{zone }4)} = \frac{(0.8)(0.4)(1.2)(1.0)(1601\text{ kN})}{5.6}$$
$$= 109.8\text{ kN}$$
$$= 109\,800\text{ N}$$

Since 620 300 N > 257 300 N $(V_{\max})$, the value of $V = 257\,300$ N for the base shear should be used. The value of $V = 257\,300$ N > $V_{\min(\text{zone }4)} = 109\,800$ N as well.

*Customary U.S. solution*

From UBC Table 16-I, $Z$ is 0.4. From UBC Table 16-K, $I$ is 1.0. From UBC Table 16-J, the soil profile type is $S_E$. From UBC Table 16-N, $R$ is 5.6. From UBC Table 16-Q, $C_a$ is $0.36N_a$. From UBC Table 16-R, $C_v$ is 0.96 $N_v$. From UBC Table 16-S, the near-source factor $N_a$ is 1.0. From UBC Table 16-T, the near-source factor $N_v$ is 1.2. From the UBC [Sec. 1630.2.2, Item1], $C_t$ is 0.020.

$$h_n = (8\text{ stories})(10\text{ ft})$$
$$= 80\text{ ft}$$
$$T = C_t(h_n)^{3/4}$$
$$= (0.020)(80\text{ ft})^{3/4}$$
$$= 0.53\text{ sec}$$

Since 0.53 sec < 0.7 sec, $F_t = 0$.

$$W = 10 \text{ k} + 20 \text{ k} + 30 \text{ k}$$
$$+ 40 \text{ k} + 50 \text{ k} + 60 \text{ k}$$
$$+ 70 \text{ k} + 80 \text{ k}$$
$$= 360 \text{ k}$$
$$C_a = 0.36 N_a$$
$$= (0.36)(1.0)$$
$$= 0.36$$
$$C_v = 0.96 N_v$$
$$= (0.96)(1.2)$$
$$= 1.15$$

Use UBC Formula 30-4.

$$V = \left(\frac{C_v I}{RT}\right) W$$
$$= \left(\frac{(1.15)(1.0)}{(5.6)(0.53)}\right)(360 \text{ k})$$
$$= 139.5 \text{ k}$$
$$= 139{,}500 \text{ lbf}$$

Based on the UBC [Sec. 1630.2.1], the total design base shear need not exceed $2.5C_a IW/R$ and should not be less than $0.11C_a IW$.

$$V_{\max} = \frac{(2.5)(0.36)(1.0)(360 \text{ k})}{5.6}$$
$$= 57.86 \text{ k}$$
$$= 57{,}860 \text{ lbf}$$

$$V_{\min} = (0.11)(0.36)(1.0)(360 \text{ k})\left(1000 \, \frac{\text{lbf}}{\text{k}}\right)$$
$$= 14{,}256 \text{ lbf}$$

Specifically for seismic zone 4, the UBC [Sec. 1630.2.1] requires that total base shear not be less than $0.8ZN_v IW/R$.

$$V_{\min(\text{zone 4})} = \frac{(0.8)(0.4)(1.2)(1.0)(360 \text{ k})}{5.6}$$
$$= 24.7 \text{ k}$$
$$= 24{,}700 \text{ lbf}$$

Since 139,500 lbf > 57,860 lbf($V_{\max}$), the value of $V = 57{,}860$ lbf for the base shear should be used. The value of $V = 57{,}860$ lbf $> V_{\min(\text{zone 4})} = 24{,}700$ lbf as well.

7. Answer D

*SI solution*

Per the UBC [Sec. 1630.5], $F_t$ is 0 as long as $T$ is 0.7 s or less. Therefore, $T_{\max} = 0.7$ s.

Use UBC Formula 30-8.

$$T = C_t(h_n)^{3/4}$$
$$0.7 = (0.0488)(h_n)^{3/4}$$
$$h_n = (14.34 \text{ m})^{3/4}$$
$$= 34.9 \text{ m}$$

*Customary U.S. solution*

Per the UBC [Sec. 1628.4], $F_t$ is 0 as long as $T$ is 0.7 sec or less. Therefore, $T_{\max} = 0.7$ sec.

Use UBC Formula 30-8.

$$T = C_t(h_n)^{3/4}$$
$$0.7 = (0.020)(h_n)^{3/4}$$
$$h_n = (35 \text{ ft})^{4/3}$$
$$= 114.5 \text{ ft}$$

8. Answer D

*SI solution*

Use UBC Formula 30-15 to distribute the base shear determined in the solution to Prob. 6 to the story levels.

$$F_x = \frac{(V - F_t)W_x h_x}{\sum_{i=1}^{n} W_i h_i}$$
$$= \frac{(220 \text{ kN} - \phi \text{kN})W_x h_x}{\sum_{i=1}^{n} W_i h_i}$$

| level | $h_x$ (m) | $W_x$ (kN) | $W_x h_x$ (kN·m) | $\dfrac{W_x h_x}{\sum W_x h_x}$ | $F_x$ (kN) |
|---|---|---|---|---|---|
| 8 | 24.4 | 44.5 | 1085.8 | 0.067 | 14.7 |
| 7 | 21.3 | 89.0 | 1895.7 | 0.117 | 25.6 |
| 6 | 18.3 | 133.5 | 2443.0 | 0.150 | 33.0 |
| 5 | 15.2 | 178.0 | 2705.6 | 0.167 | 36.7 |
| 4 | 12.2 | 222.4 | 2713.3 | 0.167 | 36.7 |
| 3 | 9.1 | 267.0 | 2429.7 | 0.150 | 33.0 |
| 2 | 6.1 | 311.0 | 1897.1 | 0.117 | 25.6 |
| 1 | 3.0 | 356.0 | 1068.0 | 0.067 | 14.7 |
| total | | 1601.4 kN | 16 238.2 kN·m | | $V = 220$ kN |

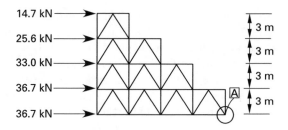

The overturning moment is the sum of the distributed shear forces at each level above the base multiplied by the moment arm measured from the pivot point at A. The overturning moment at point A is

$$(14.7 \text{ kN})(12.2 \text{ m}) + (25.6 \text{ kN})(9.1 \text{ m})$$
$$+ (33.0 \text{ kN})(6.1 \text{ m}) + (36.7 \text{ kN})(3.0 \text{ m})$$
$$\approx 724 \text{ kN·m}$$
$$\approx 723\,700 \text{ N·m}$$

*Customary U.S. solution*

Use UBC Formula 30-15 to distribute the base shear determined in the solution to Prob. 6 to the story levels.

$$F_x = \frac{(V - F_t)W_x h_x}{\sum_{i=1}^{n} W_i h_i}$$
$$= \frac{(49.5 \text{ k} - 0 \text{ k})W_x h_x}{\sum_{i=1}^{n} W_i h_i}$$

| level | $h_x$ (ft) | $W_x$ (k) | $W_x h_x$ (ft-k) | $\dfrac{W_x h_x}{\sum W_x h_x}$ | $F_x$ (k) |
|---|---|---|---|---|---|
| 8 | 80 | 10 | 800 | 0.067 | 3.31 |
| 7 | 70 | 20 | 1400 | 0.117 | 5.76 |
| 6 | 60 | 30 | 1800 | 0.150 | 7.42 |
| 5 | 50 | 40 | 2000 | 0.167 | 8.26 |
| 4 | 40 | 50 | 2000 | 0.167 | 8.26 |
| 3 | 30 | 60 | 1800 | 0.150 | 7.42 |
| 2 | 20 | 70 | 1400 | 0.117 | 5.76 |
| 1 | 10 | 80 | 800 | 0.067 | 3.31 |
| total | | 360 k | 12,000 ft-k | | $V = 49.5$ k |

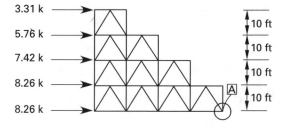

The overturning moment is the sum of the distributed shear forces at each level above the base multiplied by the moment arm measured from the pivot point at A. The overturning moment at point A is

$$(3.31 \text{ k})(40 \text{ ft}) + (5.76 \text{ k})(30 \text{ ft}) + (7.42 \text{ k})(20 \text{ ft})$$
$$+ (8.26 \text{ k})(10 \text{ ft})$$
$$= 536.2 \text{ ft-k}$$
$$= 536,000 \text{ ft-lbf}$$

**9. Answer C**

*SI solution*

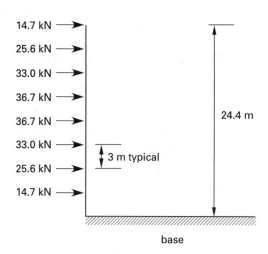

The overturning moment is the sum of the distributed shear forces at each level above the base multiplied by the moment arm measured from the pivot point at O. Use the distributed shears determined in the solution to Prob. 8. The overturning moment at the base is

$$(14.7 \text{ kN})(24.4 \text{ m}) + (25.6 \text{ kN})(21.3 \text{ m})$$
$$+ (33.0 \text{ kN})(18.3 \text{ m}) + (36.7 \text{ kN})(15.2 \text{ m})$$
$$+ (36.7 \text{ kN})(12.2 \text{ m}) + (33.0 \text{ kN})(9.1 \text{ m})$$
$$+ (25.6 \text{ kN})(6.1 \text{ m})(14.7 \text{ kN})(3.0 \text{ m})$$
$$= 3014 \text{ kN·m}$$

*Customary U.S. solution*

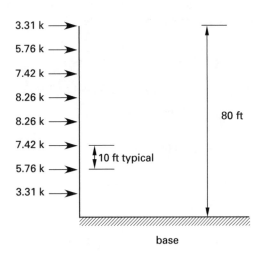

base

The overturning moment is the sum of the distributed shear forces at each level above the base multiplied by the moment arm measured from the pivot point at O. Use the distributed shears determined in the solution to Prob. 8. The overturning moment at the base is

$$(3.31 \text{ k})(80 \text{ ft}) + (5.76 \text{ k})(70 \text{ ft})$$
$$+ (7.42 \text{ k})(60 \text{ ft}) + (8.26 \text{ k})(50 \text{ ft})$$
$$+ (8.26 \text{ k})(40 \text{ ft}) + (7.42 \text{ k})(30 \text{ ft})$$
$$+ (5.76 \text{ k})(20 \text{ ft}) + (3.31 \text{ k})(10 \text{ ft})$$
$$= 2227.5 \text{ ft-k}$$

10.  Answer B

*SI solution*

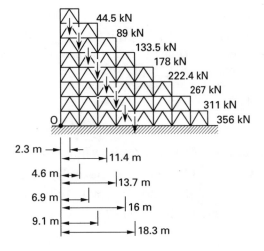

The overturning moment is resisted by the dead loads of the structure. The moment arm is shown in the figure, measured at each level from point O at the base. To reduce uplift, the UBC [Sec. 1612.3.1] limits the dead load contribution to 90%. For materials that use working stress procedures (ASD), dead load should be multiplied by a factor of 0.9. Thus, the design uplift is

$$\Big( (44.5 \text{ kN})(2.3 \text{ m}) + (89 \text{ kN})(4.6 \text{ m})$$
$$+ (133.5 \text{ kN})(6.9 \text{ m}) + (178 \text{ kN})(9.1 \text{ m})$$
$$+ (222.4 \text{ kN})(11.4 \text{ m}) + (267 \text{ kN})(13.7 \text{ m})$$
$$+ (311 \text{ kN})(16.0 \text{ m}) + (356 \text{ kN})(18.3) \Big) (0.9)$$
$$= 18\,663 \text{ kN·m}$$

*Customary U.S. solution*

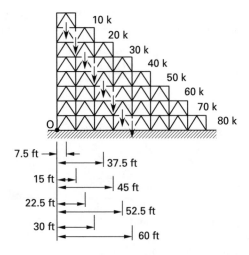

The overturning moment is resisted by the dead loads of the structure. The moment arm is shown in the figure, measured at each level from point O at the base. To reduce uplift, the UBC [Sec. 1612.3.1] limits the dead load contribution to 90%. For materials that use working stress procedures (ASD), dead load should be multiplied by a factor of 0.9. Thus, the design uplift is

$$\Big( (10 \text{ k})(7.5 \text{ ft}) + (20 \text{ k})(15.0 \text{ ft}) + (30 \text{ k})(22.5 \text{ ft})$$
$$+ (40 \text{ k})(30.0 \text{ ft}) + (50 \text{ k})(37.5 \text{ ft}) + (60 \text{ k})(45.0 \text{ ft})$$
$$+ (70 \text{ k})(52.5 \text{ ft}) + (80 \text{ k})(60.0 \text{ ft}) \Big) (0.9)$$
$$= 13,770 \text{ ft-k}$$

# C.

11. Answer C

*SI solution*

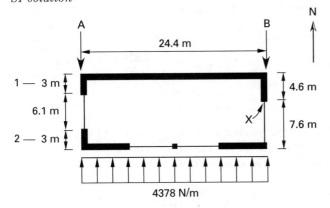

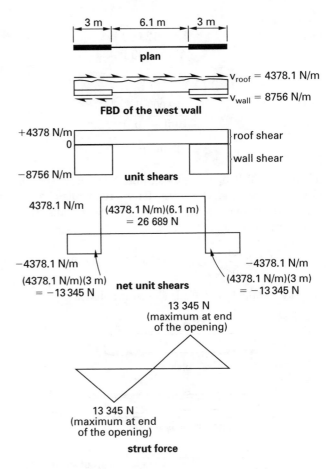

The walls along line A will resist earthquake shear forces in north-south loading.

$$V = \frac{wL}{2}$$

$$= \frac{\left(4378 \; \dfrac{\text{N}}{\text{m}}\right)(24.4 \; \text{m})}{2}$$

$$= 53\,378.4 \; \text{N}$$

At line A,

$$\vartheta_{\text{roof}} = \frac{V}{b} = \frac{53\,378.4 \; \text{N}}{12.2 \; \text{m}}$$

$$= 4378.1 \; \frac{\text{N}}{\text{m}}$$

$$\vartheta_{\text{wall}} = \frac{53\,378.4 \; \text{N}}{(3 \; \text{m} + 3 \; \text{m})}$$

$$= 8756.3 \; \text{N/m}$$

There is a 6.1 m opening along line A in the west wall. Over the opening, the drag strut force transmits the unsupported diaphragm shear to the shear walls. The drag strut force is

$$\left(8756.3 \; \frac{\text{N}}{\text{m}} - 4378.1 \; \frac{\text{N}}{\text{m}}\right)(6.1 \; \text{m}) = 26\,689 \; \text{N}$$

A more detailed analysis is shown graphically.

*Customary U.S. solution*

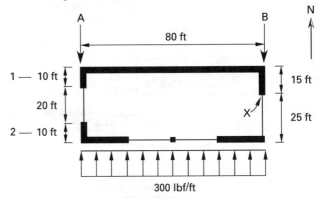

The walls along line A will resist earthquake shear forces in north-south loading.

$$V = \frac{wL}{2}$$

$$= \frac{\left(300 \; \dfrac{\text{lbf}}{\text{ft}}\right)(80 \; \text{ft})}{2}$$

$$= 12{,}000 \; \text{lbf}$$

At line A,

$$\vartheta_{\text{roof}} = \frac{V}{b} = \frac{12{,}000 \text{ lbf}}{40 \text{ ft}}$$

$$= 300 \text{ lbf/ft}$$

$$\vartheta_{\text{wall}} = \frac{12{,}000 \text{ lbf}}{10 \text{ ft} + 10 \text{ ft}}$$

$$= 600 \text{ lbf/ft}$$

There is a 20 ft opening along line A in the west wall. Over the opening, the drag strut force transmits the unsupported diaphragm shear to the shear walls. The drag strut force is

$$\left(600 \, \frac{\text{lbf}}{\text{ft}} - 300 \, \frac{\text{lbf}}{\text{ft}}\right)(20 \text{ ft}) = 6000 \text{ lbf}$$

A more detailed analysis is shown graphically.

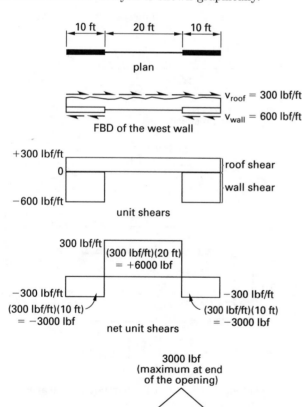

## 12. Answer D

*SI solution*

The wall along line B will resist earthquake shear forces in north-south loading. From the solution to Prob. 11,

$$V_{\text{shear}} = 53\,378 \text{ N}$$

$$\vartheta = \frac{V}{b} = \frac{53\,378.4 \text{ N}}{12.2 \text{ m}}$$

$$= 4378.1 \text{ N/m}$$

The unsupported diaphragm shear over the opening will be transmitted by the drag strut to the shear wall at point X. The drag strut force is

$$\left(4378.1 \, \frac{\text{N}}{\text{m}}\right)(7.6 \text{ m}) = 33\,361 \text{ N}$$

*Customary U.S. solution*

The wall along line B will resist earthquake shear forces in north-south loading. From the solution to Prob. 11,

$$V_{\text{shear}} = 12{,}000 \text{ lbf}$$

$$\vartheta = \frac{V}{b} = \frac{12{,}000 \text{ lbf}}{40 \text{ ft}}$$

$$= 300 \text{ lbf/ft}$$

The unsupported diaphragm shear over the opening will be transmitted by the drag strut to the shear wall at point X. The drag strut force is

$$\left(300 \, \frac{\text{lbf}}{\text{ft}}\right)(25 \text{ ft}) = 7500 \text{ lbf}$$

## 13. Answer A

*SI solution*

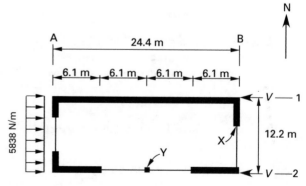

The walls along line 2 will resist earthquake shear forces in east-west loading.

$$V = \frac{wL}{2} = \frac{\left(5838 \ \dfrac{\text{N}}{\text{m}}\right)(12.2 \text{ m})}{2}$$

$$= 35\,585.6 \text{ N}$$

$$\vartheta_{\text{roof}} = \frac{V}{b} = \frac{35\,585.6 \text{ N}}{24.4 \text{ m}}$$

$$= 1459.4 \text{ N/m}$$

$$\vartheta_{\text{wall}} = \frac{35\,586.5 \text{ N}}{12.2 \text{ m}}$$

$$= 2918.8 \text{ N/m}$$

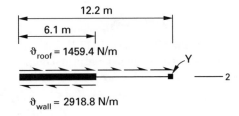

The drag strut force at Y is

$$\left(1459.4 \ \frac{\text{N}}{\text{m}}\right)(12.2 \text{ m}) - \left(2918.8 \ \frac{\text{N}}{\text{m}}\right)(6.1 \text{ m}) = 0 \text{ N}$$

*Customary U.S. solution*

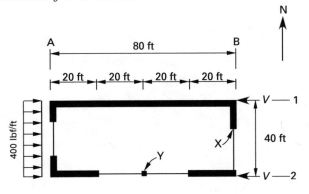

The walls along line 2 will resist earthquake shear forces in east-west loading.

$$V = \frac{wL}{2} = \frac{\left(400 \ \dfrac{\text{lbf}}{\text{ft}}\right)(40 \text{ ft})}{2}$$

$$= 8000 \text{ lbf}$$

At line 2,

$$\vartheta_{\text{roof}} = \frac{V}{b} = \frac{8000 \text{ lbf}}{80 \text{ ft}}$$

$$= 100 \text{ lbf/ft}$$

$$\vartheta_{\text{wall}} = \frac{8000 \text{ lbf}}{40 \text{ ft}}$$

$$= 200 \text{ lbf/ft}$$

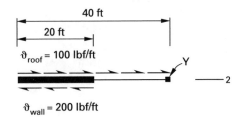

The drag strut force at Y is

$$\left(100 \ \frac{\text{lbf}}{\text{ft}}\right)(40 \text{ ft}) - \left(200 \ \frac{\text{lbf}}{\text{ft}}\right)(20 \text{ ft}) = 0 \text{ lbf}$$

**14.  Answer B**

*SI solution*

The north-south loading should be considered. The shear force in the wall along line B is as calculated in the solution to Prob. 11.

$$V = 53\,378.4 \text{ N}$$

$$\vartheta_{\text{wall}} = \frac{53\,378.4 \text{ N}}{4.6 \text{ m}}$$

$$= 11\,675 \text{ N/m}$$

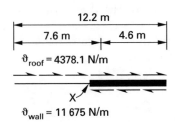

At point X,

$$-D_{\text{X}} + \left(4378.1 \ \frac{\text{N}}{\text{m}}\right)(7.6 \text{ m}) = 0$$

The drag strut force, $D_X$, is

$$D_X = 33\,361 \text{ N} \quad \text{[tension]}$$

*Customary U.S. solution*

The north-south loading should be considered. The shear force in the wall along line B is as calculated in the solution to Prob. 11.

$$V = 12{,}000 \text{ lbf}$$

$$\vartheta_{\text{wall}} = \frac{12{,}000 \text{ lbf}}{15 \text{ ft}}$$

$$= 800 \text{ lbf/ft}$$

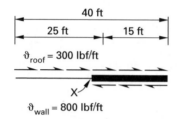

At point X,

$$-D_X + \left(300 \, \frac{\text{lbf}}{\text{ft}}\right)(25 \text{ ft}) = 0$$

The drag strut force, $D_X$, is

$$D_X = 7500 \text{ lbf} \quad \text{[tension]}$$

**15. Answer B**
The magnitude of the drag strut force at Y depends on the east-west loading, not the north-south loading. Therefore, the drag strut force remains unchanged when the north-south loading increases.

# D.

**16. Answer C**
*SI solution*

According to the UBC [Sec. 1632.1], lateral force on elements of structures, nonstructural components, and equipment supported by structures should be determined from UBC Formula 32-1, $F_p = 4.0C_aI_pW_p$.

From UBC Table 16-I, $Z$ is 0.4. From UBC Table 16-K, $I_p$ is 1.0. From UBC Table 16-J, the soil profile type is $S_C$ for a very dense soil. From UBC Table 16-Q, the seismic coefficient $C_a = 0.40N_a$. From UBC Table 16-S, near-source factor $N_a$ is 1.0. Therefore,

$$C_a = (0.40)(1.0)$$

$$= 0.40$$

Use UBC Formula 32-1.

$$V = (4.0)(0.40)(1.0)(133\,446 \text{ N})$$

$$= 213\,514 \text{ N}$$

*Customary U.S. solution*

According to the UBC [Sec. 1632.1], lateral force on elements of structures, nonstructural components, and equipment supported by structures should be determined from UBC Formula 32-1, $F_p = 4.0C_aI_pW_p$.

From UBC Table 16-I, $Z$ is 0.4. From UBC Table 16-K, $I_p$ is 1.0. From UBC Table 16-J, the soil profile type is $S_C$ for a very dense soil. From UBC Table 16-Q, the seismic coefficient $C_a = 0.40N_a$. From UBC Table 16-S, near-source factor $N_a$ is 1.0. Therefore,

$$C_a = (0.40)(1.0)$$

$$= 0.40$$

Use UBC Formula 32-1.

$$V = (4.0)(0.40)(1.0)(30{,}000 \text{ lbf})$$

$$= 48{,}000 \text{ lbf}$$

**17. Answer B**
*SI solution*

Based on the UBC [Sec. 1634.1.1], this tank is a non-building structure. Since the fundamental period of 0.05 s is less than 0.06 s, tank II is also a rigid structure. According to the UBC [Sec. 1634.4], tanks with flat bottoms at or below grade should be designed for the lateral force obtained from UBC Formula 34-1, $V = 0.7C_aIW$.

From UBC Table 16-I, $Z$ is 0.4. From UBC Table 16-K, $I$ is 1.0. From UBC Table 16-J, the soil profile type is $S_A$ for a hard rock material. From UBC Table 16-Q, the seismic response coefficient $C_a = 0.32N_a$. From UBC Table 16-S, near-source factor $N_a$ is 1.0. Therefore,

$$C_a = (0.32)(1.0)$$
$$= 0.32$$

Use UBC Formula 34-1.

$$V = (0.7)(0.32)(1.0)(444\,820\text{ N})$$
$$= 99\,640\text{ N}$$

*Customary U.S. solution*

Based on the UBC [Sec. 1634.1.1], this tank is a non-building structure. Since the fundamental period of 0.05 sec is less than 0.06 sec, tank II is also a rigid structure. According to the UBC [Sec. 1634.4], tanks with flat bottoms at or below grade should be designed for the lateral force obtained from UBC Formula 34-1, $V = 0.7C_aIW$.

From UBC Table 16-I, $Z$ is 0.4. From UBC Table 16-K, $I$ is 1.0. From UBC Table 16-J, the soil profile type is $S_A$ for a hard rock material. From UBC Table 16-Q, the seismic response coefficient $C_a = 0.32N_a$. From UBC Table 16-S, near-source factor $N_a$ is 1.0. Therefore,

$$C_a = (0.32)(1.0)$$
$$= 0.32$$

Use UBC Formula 34-1.

$$V = (0.7)(0.32)(1.0)(100{,}000\text{ lbf})$$
$$= 22{,}400\text{ lbf}$$

18. Answer B

*SI solution*

The guyed tower stack is a nonbuilding structure. Based on the UBC [Sec. 1634.1.1], nonbuilding structures include all self-supporting structures other than buildings. They carry gravity loads and resist the effects of earthquakes. Since the fundamental period of 0.01 s < 0.06 s, a trussed tower is considered to be a rigid structure.

Based on the UBC [Sec. 1634.3], nonbuilding rigid structures and their anchorages should be designed for the lateral force determined from UBC Formula 34-1, $V = 0.7C_aIW$.

From UBC Table 16-I, $Z$ is 0.4. From UBC Table 16-K, $I$ is 1.0. From UBC Table 16-J, the soil profile type is

$S_E$ for a soft clay soil. From UBC Table 16-Q, the seismic response coefficient $C_a = 0.36N_a$. From UBC Table 16-S, near-source factor $N_a$ is 1.0. Therefore,

$$C_a = (0.36)(1.0)$$
$$= 0.36$$

Use UBC Formula 34-1.

$$V = (0.7)(0.36)(1.0)(66\,723\text{ N})$$
$$= 16\,814\text{ N}$$

*Customary U.S. solution*

The guyed tower stack is a nonbuilding structure. Based on the UBC [Sec. 1634.1.1], nonbuilding structures include all self-supporting structures other than buildings. They carry gravity loads and resist the effects of earthquakes. Since the fundamental period of 0.01 sec < 0.06 sec, a trussed tower is considered to be a rigid structure.

Based on the UBC [Sec. 1634.3], nonbuilding rigid structures and their anchorages should be designed for the lateral force determined from UBC Formula 34-1, $V = 0.7C_aIW$.

From UBC Table 16-I, $Z$ is 0.4. From UBC Table 16-K, $I$ is 1.0. From UBC Table 16-J, the soil profile type is $S_E$ for a soft clay soil. From UBC Table 16-Q, the seismic response coefficient $C_a = 0.36N_a$. From UBC Table 16-S, near-source factor $N_a$ is 1.0. Therefore,

$$C_a = (0.36)(1.0)$$
$$= 0.36$$

Use UBC Formula 34-1.

$$V = (0.7)(0.36)(1.0)(15{,}000\text{ lbf})$$
$$= 3780\text{ lbf}$$

19. Answer C

*SI solution*

Based on the UBC [Sec. 1634.3], the guyed tower stack is a nonrigid nonbuilding structure, because the fundamental period of 0.12 s > 0.06 s.

According to the UBC [Sec. 1634.5], nonbuilding structures that are not covered by sections 1634.3 (rigid structures) and 1634.4 (tanks with supported bottoms) should be designed to resist seismic forces determined by UBC Formula 30-4.

$$V = \left(\frac{C_v I}{RT}\right) W$$

From UBC Table 16-I, $Z$ is 0.4. From UBC Table 16-K, $I$ is 1.0. From UBC Table 16-J, the soil profile type is $S_E$ for a soft clay soil. From UBC Table 16-P, $R$ is 2.9. From UBC Table 16-R, the seismic response coefficient $C_v = 0.96 N_v$. From UBC Table 16-T, the near-source factor $N_v$ is 1.0. Therefore,

$$C_v = (0.96)(1.0)$$
$$= 0.96$$

Use UBC Formula 30-4.

$$V = \left(\frac{(0.96)(1.0)}{(2.9)(0.12)}\right)(66\,723 \text{ N})$$
$$= 184\,063 \text{ N}$$

Per the UBC [Sec. 1634.5], the total design base shear should not be less than UBC Formula 34-2, $V = 0.56 C_a I W$, nor less than UBC Formula 34-3, $V = 1.6 Z N_v I W / R$ for seismic zone 4.

From UBC Table 16-Q, the seismic response coefficient $C_a = 0.36 N_a$. From UBC Table 16-S, the near-source factor $N_a$ is 1.0. Therefore,

$$C_a = (0.36)(1.0)$$
$$= 0.36$$

Use UBC Formula 34-2.

$$V = (0.56)(0.36)(1.0)(66\,723 \text{ N})$$
$$= 13\,451 \text{ N}$$

Use UBC Formula 34-3.

$$V = \frac{(1.6)(0.4)(1.0)(1.0)(66\,723 \text{ N})}{2.9}$$
$$= 14\,725 \text{ N}$$

Since $V = 184\,063$ N is greater than $V = 13\,451$ N and $V = 14\,725$ N, $V = 184\,063$ N can be used. However, based on the UBC [Sec. 1630.2.1], the total design base shear need not exceed UBC Formula 30-5.

$$V_{\max} = \left(\frac{2.5 C_a I}{R}\right) W$$
$$= \left(\frac{(2.5)(0.36)(1.0)}{2.9}\right)(66\,723 \text{ N})$$
$$= 20\,707 \text{ N}$$

$V = 184\,063$ N is greater than $V_{\max} = 20\,707$ N. Therefore, $V_{\max} = 20\,707$ N becomes the controlling value.

*Customary U.S. solution*

Based on the UBC [Sec. 1634.3], the guyed tower stack is a nonrigid nonbuilding structure, because the fundamental period of 0.12 sec > 0.06 sec.

According to the UBC [Sec. 1634.5], nonbuilding structures not covered by sections 1634.3 (rigid structures) and 1634.4 (tanks with supported bottoms) should be designed to resist seismic forces determined by UBC Formula 30-4.

$$V = \left(\frac{C_v I}{RT}\right) W$$

From UBC Table 16-I, $Z$ is 0.4. From UBC Table 16-K, $I$ is 1.0. From UBC Table 16-J, the soil profile type is $S_E$ for a soft clay soil. From UBC Table 16-P, $R$ is 2.9. From UBC Table 16-R, the seismic response coefficient $C_v = 0.96 N_v$. From UBC Table 16-T, the near-source factor $N_v$ is 1.0. Therefore,

$$C_v = (0.96)(1.0)$$
$$= 0.96$$

Use UBC Formula 30-4.

$$V = \left(\frac{(0.96)(1.0)}{(2.9)(0.12)}\right)(15,000 \text{ lbf})$$
$$= 41,379 \text{ lbf}$$

Per the UBC [Sec. 1634.5], the total design base shear should not be less than UBC Formula 34-2, $V = 0.56 C_a I W$, nor less than UBC Formula 34-3, $V = 1.6 Z N_v I W / R$ for seismic zone 4.

From UBC Table 16-Q, the seismic response coefficient $C_a = 0.36 N_a$. From UBC Table 16-S, the near-source factor $N_a$ is 1.0. Therefore,

$$C_a = (0.36)(1.0)$$
$$= 0.36$$

Use UBC Formula 34-2.

$$V = (0.56)(0.36)(1.0)(15,000 \text{ lbf})$$
$$= 3024 \text{ lbf}$$

Use UBC Formula 34-3.

$$V = \frac{(1.6)(0.4)(1.0)(1.0)(15,000 \text{ lbf})}{2.9}$$
$$= 3310 \text{ lbf}$$

Since $V = 41{,}379$ lbf is greater than $V = 3024$ lbf and $V = 3310$ lbf, $V = 41{,}379$ lbf can be used. However, based on the UBC [Sec. 1630.2.1], the total design base shear need not exceed UBC Formula 30-5.

$$
\begin{aligned}
V_{\max} &= \left(\frac{2.5 C_a I}{R}\right) W \\
&= \frac{(2.5)(0.36)(1.0)(15{,}000 \text{ lbf})}{2.9} \\
&= 4655 \text{ lbf}
\end{aligned}
$$

$V = 41{,}379$ lbf is greater than $V_{\max} = 4655$ lbf. Therefore, $V_{\max} = 4655$ lbf becomes the controlling value.

20. Answer B

*SI solution*

Since the fundamental period of 0.04 s is less than 0.06 s, tank II (nonbuilding structure) is a rigid structure. Based on the UBC [Secs. 1634.3 and 1634.4], for tanks with flat bottoms at or below grade, UBC Formula 34-1, $V = 0.7 C_a I_W$, should be used to obtain the lateral force.

From UBC Table 16-I, $Z$ is 0.4. From UBC Table 16-K, for nonbuilding structures housing, supporting, or containing quantities of toxic or explosive substances that would be hazardous to the safety of the general public if released, the value of $I$ should be taken as 1.25. From UBC Table 16-J, $S_A$ can be assigned for the soil profile type. From UBC Table 16-Q, the seismic response coefficient $C_a = 0.32 N_a$. From UBC Table 16-S, near-source factor $N_a$ is 1.0. Therefore,

$$
\begin{aligned}
C_a &= (0.32)(1.0) \\
&= 0.32
\end{aligned}
$$

Use UBC Formula 34-1.

$$
\begin{aligned}
V &= (0.7)(0.32)(1.25)(444\,820 \text{ N}) \\
&= 124\,550 \text{ N}
\end{aligned}
$$

*Customary U.S. solution*

Since the fundamental period of 0.04 sec is less than 0.06 sec, tank II (nonbuilding structure) is a rigid structure. Based on the UBC [Secs. 1634.3 and 1634.4], for tanks with flat bottoms at or below grade, UBC Formula 34-1, $V = 0.7 C_a I W$, should be used to obtain the lateral force.

From UBC Table 16-I, $Z$ is 0.4. From UBC Table 16-K, for nonbuilding structures housing, supporting, or containing quantities of toxic or explosive substances that would be hazardous to the safety of the general public if released, the value of $I$ should be taken as 1.25. From UBC Table 16-J, $S_A$ can be assigned for the soil profile type. From UBC Table 16-Q, the seismic response coefficient $C_a = 0.32 N_a$. From UBC Table 16-S, near-source factor $N_a$ is 1.0. Therefore,

$$
\begin{aligned}
C_a &= (0.32)(1.0) \\
&= 0.32
\end{aligned}
$$

Use UBC Formula 34-1.

$$
\begin{aligned}
V &= (0.7)(0.32)(1.25)(100{,}000 \text{ lbf}) \\
&= 28{,}000 \text{ lbf}
\end{aligned}
$$

# E.

21. Answer C

The connections must be designed to resist the acceleration of the mass of the wall during the earthquake. Use UBC Formula 32-1. This formula specifies the total design lateral force on elements of structures—in this case, the perpendicular wall.

$$
F_p = 4.0 C_a I_p W_p
$$

22. Answer B

*SI solution*

From UBC Table 16-I, $Z$ is 0.4. From UBC Table 16-N, $R$ is 4.5. Based on the UBC [Sec. 1630.2.3.2], type $S_D$ soil profile should be used in seismic zone 4 since the soil properties are not known. From UBC Table 16-Q, the seismic response coefficient $C_a$ is $0.44 N_a$. From UBC Table 16-S, the near-source factor $N_a$ is 1.2. Therefore,

$$
\begin{aligned}
C_a &= 0.44 N_a \\
&= (0.44)(1.2) \\
&= 0.53
\end{aligned}
$$

Note that based on the UBC [Sec. 1630.2.3.2], the value of the near-source factor need not be greater than 1.3 in seismic zone 4 if structural irregularities of type 1, 4, or 5 of UBC Table 16-L or type 1 or 4 of UBC Table 16-M are not present.

$$D_{\text{roof}} = (36.6 \text{ m})(18.3 \text{ m})\left(958 \; \frac{\text{N}}{\text{m}^2}\right)$$

$$= 641\,649 \text{ N}$$

$$D_{\text{walls}} = (2 \text{ walls})(36.6 \text{ m})\left(\frac{5.7}{2} \text{ m} + 1 \text{ m}\right)\left(4788 \; \frac{\text{N}}{\text{m}^2}\right)$$

$$= 1\,349\,354 \text{ N}$$

$$W = D_{\text{roof}} + D_{\text{walls}}$$

$$= 641\,649 \text{ N} + 1\,349\,354 \text{ N}$$

$$= 1\,991\,000 \text{ N}$$

Use UBC Formula 30-11.

$$V = \left(\frac{3.0 C_a}{R}\right) W$$

$$= \left(\frac{(3.0)(0.53)}{4.5}\right)(1\,991\,000 \text{ N})$$

$$= 703\,487 \text{ N}$$

The seismic loading to the roof diaphragm is

$$w_{\text{N-S}} = \frac{703\,487 \text{ N}}{36.6 \text{ m}}$$

$$= 19\,221 \text{ N/m}$$

$$\approx 19\,220 \text{ N/m}$$

*Customary U.S. solution*

From UBC Table 16-I, $Z$ is 0.4. From UBC Table 16-N, $R$ is 4.5. Based on the UBC [Sec. 1630.2.3.2], type $S_D$ soil profile should be used in seismic zone 4, since the soil properties are not known. From UBC Table 16-Q, the seismic response coefficient $C_a$ is $0.44 N_a$. From UBC Table 16-S, the near-source factor $N_a$ is 1.2. Therefore,

$$C_a = 0.44 N_a$$

$$= (0.44)(1.2)$$

$$= 0.53$$

Note that based on the UBC [Sec. 1630.2.3.2], the value of the near-source factor need not be greater than 1.3 in seismic zone 4 if structural irregularities of type 1, 4, or 5 of UBC Table 16-L or type 1 or 4 of UBC Table 16-M are not present.

$$D_{\text{roof}} = (120 \text{ ft})(60 \text{ ft})\left(20 \; \frac{\text{lbf}}{\text{ft}^2}\right)$$

$$= 144{,}000 \text{ lbf}$$

$$D_{\text{walls}} = (2 \text{ walls})(120 \text{ ft})\left(\frac{18.67}{2} \text{ ft} + 3.34 \text{ ft}\right)\left(100 \; \frac{\text{lbf}}{\text{ft}^2}\right)$$

$$= 304{,}200 \text{ lbf}$$

$$W = D_{\text{roof}} + D_{\text{walls}}$$

$$= 144{,}000 \text{ lbf} + 304{,}200 \text{ lbf}$$

$$= 448{,}200 \text{ lbf}$$

Use UBC Formula 30-11.

$$V = \left(\frac{3.0 C_a}{R}\right) W$$

$$= \left(\frac{(3.0)(0.53)}{4.5}\right)(448{,}200 \text{ lbf})$$

$$= 158{,}364 \text{ lbf}$$

The seismic loading to the roof diaphragm is

$$w_{\text{N-S}} = \frac{158{,}364 \text{ lbf}}{120 \text{ ft}}$$

$$= 1320 \text{ lbf/ft}$$

**23.** Answer C

*SI solution*

Based on the UBC [Sec. 1632], UBC Formula 32-2 should be used. From UBC Table 16-I, $Z$ is 0.4. From UBC Table 16-K, $I_p$ is 1.0. Per the UBC [Sec. 1630.2.3.2], the soil profile type is $S_D$ because sufficient details regarding the soil properties are not available. From UBC Table 16-Q, the seismic response coefficient $C_a$ is $0.44 N_a$. From UBC Table 16-S, the near-source factor $N_a$ is 1.2. Therefore,

$$C_a = 0.44 N_a$$

$$= (0.44)(1.2)$$

$$= 0.53$$

From UBC Table 16-O, the values of $a_p$ and $R_p$ are 1.0 and 3.0, respectively.

$$W_p = D(h)$$

Note that only the top half of the wall contributes its weight to the inertia forces; the weight of the bottom half is transferred to the foundation.

For the main wall,

$$W_p = \left(4788 \ \frac{\text{N}}{\text{m}^2}\right)\left(\frac{5.7 \text{ m}}{2}\right)$$
$$= 13\,646 \text{ N/m}$$

For the parapet wall,

$$W_p = \left(4788 \ \frac{\text{N}}{\text{m}^2}\right)(1 \text{ m})$$
$$= 4788 \text{ N/m}$$

Use UBC Formula 32-2.

$$F_p = \left(\frac{a_p C_a I_p}{R_p}\right)\left(1 + (3)\left(\frac{h_x}{h_r}\right)\right)W_p$$

Per the UBC [Sec. 1632.2], $h_x$ is the element or component *attachment elevation* with respect to grade, and $h_r$ is the structure's *roof elevation* with respect to grade.

For the main wall,

$$F_p = \left(\frac{(1.0)(0.53)(1.0)}{3.0}\right)$$
$$\times \left(1 + (3)\left(\frac{5.7 \text{ m}}{5.7 \text{ m}}\right)\right)\left(13\,646 \ \frac{\text{N}}{\text{m}}\right)$$
$$= 9643 \text{ N/m}$$

For the parapet wall,

$$F_p = \left(\frac{(1.0)(0.53)(1.0)}{3.0}\right)$$
$$\times \left(1 + (3)\left(\frac{5.7 \text{ m}}{5.7 \text{ m}}\right)\right)\left(4788 \ \frac{\text{N}}{\text{m}}\right)$$
$$= 3383.5 \text{ N/m}$$

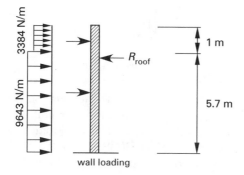

wall loading

The UBC [Secs. 1605.2.3, 1611.4, and 1633.2.8.1] states that concrete or masonry walls should be anchored to all floors, roofs, and other structural elements that provide required lateral support for the wall. Such anchorages should provide a positive direct connection capable of resisting the horizontal forces specified in UBC Chapter 16 or a minimum force of 4.09 kN/m of wall, or in zone 4 a minimum force of 6.1 kN/m of wall, whichever is greater.

$$\Sigma M_{\text{base}} = 0$$

$$\left(9643 \ \frac{\text{N}}{\text{m}}\right)\left(\frac{5.7 \text{ m}}{2}\right)$$
$$+ \left(3383.5 \ \frac{\text{N}}{\text{m}}\right)\left(\frac{1 \text{ m}}{2} + 5.7 \text{ m}\right) - R_{\text{roof}}(5.7 \text{ m}) = 0$$
$$R_{\text{roof}} = 8501.8 \text{ N} \quad [\text{per meter of wall}]$$
$$= 8.5 \text{ kN/m}$$

The calculated force of 8.5 kN/m is greater than the minimum requirement of 6.1 kN/m of wall. Since anchorage points are at 1.2 m O.C.,

$$F_{\text{anchorage}} = \left(8501.8 \ \frac{\text{N}}{\text{m}}\right)(1.2 \text{ m})$$
$$= 10\,202 \text{ N}$$

*Customary U.S. solution*

Based on the UBC [Sec. 1632], UBC Formula 32-2 should be used. From UBC Table 16-I, $Z$ is 0.4. From UBC Table 16-K, $I_p$ is 1.0. Per the UBC [Sec. 1630.2.3.2], the soil profile type is $S_D$ because sufficient details regarding the soil properties are not available. From UBC Table 16-Q, the seismic response coefficient $C_a$ is $0.44N_a$. From UBC Table 16-S, the near-source factor $N_a$ is 1.2. Therefore,

$$C_a = 0.44N_a$$
$$= (0.44)(1.2)$$
$$= 0.53$$

From UBC Table 16-O, the values of $a_p$ and $R_p$ are 1.0 and 3.0, respectively.

$$W_p = D(h)$$

Note that only the top half of the wall contributes its weight to the inertia force; the weight of the bottom half is transferred to the foundation.

For the main wall,

$$W_p = \left(100 \ \frac{\text{lbf}}{\text{ft}^2}\right)\left(\frac{18.67 \ \text{ft}}{2}\right)$$
$$= 933.5 \ \text{lbf/ft}$$

For the parapet wall,

$$W_p = \left(100 \ \frac{\text{ft}}{\text{ft}^2}\right)(3.34 \ \text{ft})$$
$$= 334 \ \text{lbf/ft}$$

Use UBC Formula 32-2.

$$F_p = \left(\frac{a_p C_a I_p}{R_p}\right)\left(1 + (3)\left(\frac{h_x}{h_r}\right)\right)W_p$$

Per the UBC [Sec. 1632.2], $h_x$ is the element or component *attachment elevation* with respect to grade, and $h_r$ is the structure's *roof elevation* with respect to grade.

For the main wall,

$$F_p = \left(\frac{(1.0)(0.53)(1.0)}{3.0}\right)$$
$$\times \left(1 + (3)\left(\frac{18.67 \ \text{ft}}{18.67 \ \text{ft}}\right)\right)\left(933.5 \ \frac{\text{lbf}}{\text{ft}}\right)$$
$$= 659.7 \ \text{lbf/ft}$$

For the parapet wall,

$$F_p = \left(\frac{(1.0)(0.53)(1.0)}{3.0}\right)$$
$$\times \left(1 + (3)\left(\frac{18.67 \ \text{ft}}{18.67 \ \text{ft}}\right)\right)\left(334 \ \frac{\text{lbf}}{\text{ft}}\right)$$
$$= 236.0 \ \text{lbf/ft}$$

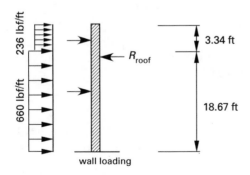

wall loading

The UBC [Secs. 1605.2.3, 1611.4, and 1633.2.8.1] states that concrete or masonry walls should be anchored to all floors, roofs, and other structural elements that provide required lateral support for the wall. Such anchorages should provide a positive direct connection capable of resisting the horizontal forces specified in UBC Chapter 16 or a minimum force of 280 pounds per lineal foot of wall, or in zone 4, a minimum force of 420 pounds per lineal foot of wall, whichever is greater.

$$\Sigma M_{\text{base}} = 0$$

$$\left(659.7 \ \frac{\text{lbf}}{\text{ft}}\right)\left(\frac{18.67 \ \text{ft}}{2}\right)$$
$$+ \left(236.0 \ \frac{\text{lbf}}{\text{ft}}\right)\left(\frac{3.34 \ \text{ft}}{2} + 18.67 \ \text{ft}\right)$$
$$- R_{\text{roof}}(18.67 \ \text{ft}) = 0$$
$$R_{\text{roof}} = 587.0 \ \text{lbf} \quad [\text{per foot of wall}]$$
$$= 8.5 \ \text{kN/m}$$

The calculated force of 587 pounds per lineal foot of wall is greater than the minimum requirement of 420 pounds per lineal foot. Since anchorage points are at 4 ft O.C.,

$$F_{\text{anchorage}} = (587.0 \ \text{lbf})(4 \ \text{ft})$$
$$= 2348 \ \text{lbf}$$

**24. Answer D**

*SI solution*

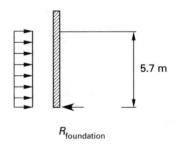

$R_{\text{foundation}}$

The seismic force resisted at the foundation is from the acceleration of the mass of the lower half of the wall. From UBC Table 16-I, $Z$ is 0.4. From UBC Table 16-K, $I_p$ is 1.0. Per the UBC [Sec. 1630.2.3.2], $S_D$ should be assigned for the soil profile type. From UBC Table

16-Q, the seismic coefficient $C_a$ is $0.44N_a$. From UBC Table 16-S, the near-source factor $N_a$ is 1.2. Therefore,

$$C_a = 0.44N_a$$
$$= (0.44)(1.2)$$
$$= 0.53$$

From UBC Table 16-O, the values of $a_p$ and $R_p$ are 1.0 and 3.0, respectively.

Based on the UBC [Sec. 1632], use UBC Formula 32-2.

$$F_p = \left(\frac{a_p C_a I_p}{R_p}\right)\left(1 + (3)\left(\frac{h_x}{h_r}\right)\right)W_p$$

Per the UBC [Sec. 1632.2], $h_x$ is the element or component *attachment elevation* with respect to grade, and $h_r$ is the structure's *roof elevation* with respect to grade.

For the main wall,

$$= \left(\frac{(1.0)(0.53)(1.0)}{3.0}\right)\left(1 + (3)\left(\frac{6.7 \text{ m}}{5.7 \text{ m}}\right)\right)$$
$$\times \left(4788 \frac{\text{N}}{\text{m}^2}\right)$$
$$= 3828.72 \text{ N/m}^2$$

$$R_{\text{foundation}} = F_p\left(\frac{h}{2}\right)$$
$$= \left(3828.72 \frac{\text{N}}{\text{m}^2}\right)\left(\frac{5.7 \text{ m}}{2}\right)$$
$$= 10\,912 \text{ N/m}$$

*Customary U.S. solution*

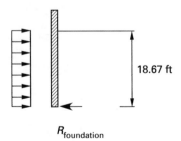

$R_{\text{foundation}}$

The seismic force resisted at the foundation is from the acceleration of the mass of the lower half of the wall. From UBC Table 16-I, $Z$ is 0.4. From UBC Table 16-K, $I_p$ is 1.0. Per the UBC [Sec. 1630.2.3.2], $S_D$ should

be assigned for the soil profile type. From UBC Table 16-Q, the seismic coefficient $C_a$ is $0.44N_a$. From UBC Table 16-S, the near-source factor $N_a$ is 1.2. Therefore,

$$C_a = 0.44N_a$$
$$= (0.44)(1.2)$$
$$= 0.53$$

From UBC Table 16-O, the values of $a_p$ and $R_p$ are 1.0 and 3.0, respectively.

Based on the UBC [Sec. 1632], use UBC Formula 32-2.

$$F_p = \left(\frac{a_p C_a I_p}{R_p}\right)\left(1 + (3)\left(\frac{h_x}{h_r}\right)\right)W_p$$

Per the UBC [Sec. 1632.2], $h_x$ is the element or component *attachment elevation* with respect to grade, and $h_r$ is the structure's *roof elevation* with respect to grade.

For the main wall,

$$= \left(\frac{(1.0)(0.53)(1.0)}{3.0}\right)\left(1 + (3)\left(\frac{22 \text{ ft}}{18.67 \text{ ft}}\right)\right)$$
$$\times \left(100 \frac{\text{lbf}}{\text{ft}^2}\right)$$
$$= 80.12 \text{ lbf/ft}^2$$

$$R_{\text{foundation}} = F_p\left(\frac{h}{2}\right)$$
$$= \left(80.12 \frac{\text{lbf}}{\text{ft}^2}\right)\left(\frac{18.67 \text{ ft}}{2}\right)$$
$$= 748 \text{ lbf/ft}$$

**25. Answer C**

*SI solution*

The force on the diaphragm per lineal meter was calculated in the solution to Prob. 22. The chord force is given by

$$C = \frac{wL^2}{8b}$$
$$= \frac{\left(19\,220 \frac{\text{N}}{\text{m}}\right)(36.6 \text{ m})^2}{(8)(18.3 \text{ m})}$$
$$= 175\,863 \text{ N}$$

The chord reinforcement area is calculated from the maximum allowable tensile stress in the steel. $f_y$ for grade 60 steel is 413.7 MPa. The UBC [Sec. 1926.3.2, Items 1 and 2] specifies the maximum tensile stress in steel reinforcement. For Grade 60 or greater reinforcement, the maximum allowable tensile stress is 165 475 kPa. From the UBC [Sec. 2107.2.11, Item 1.1], $F_s = 0.5f_y$, 165.5 MPa maximum. The UBC [Sec. 1603.5] allows a one-third increase in allowable stresses for seismic forces.

$$A_s = \frac{T}{F_s} = \frac{T}{(0.5f_y)(1.33)}$$

For grade 60, $0.5f_y$ is 206.8 MPa; however, the maximum allowable stress is 165.5 MPa. Therefore,

$$A_s = \frac{175\,863 \text{ N}}{(165.5 \text{ MPa})(1.33)\left(100\,\dfrac{\text{cm}}{\text{m}}\right)}$$

$$= 8.0 \text{ cm}^2$$

*Customary U.S. solution*

The force on the diaphragm per lineal foot was calculated in the solution to Prob. 22. The chord force is given by

$$C = \frac{wL^2}{8b} = \frac{\left(1320\,\dfrac{\text{lbf}}{\text{ft}}\right)(120 \text{ ft})^2}{(8)(60 \text{ ft})}$$

$$= 39{,}600 \text{ lbf}$$

The chord reinforcement area is calculated from the maximum allowable tensile stress in the steel. $f_y$ for grade 60 steel is 60,000 lbf/in$^2$. The UBC [Sec. 1926.3.2, Items 1 and 2] specifies the maximum tensile stress in steel reinforcement. For Grade 60 or greater reinforcement, the maximum allowable tensile stress is 24,000 lbf/in$^2$. From UBC [Sec. 2107.2.11, Item 1.1], $F_s = 0.5f_y$, 24,000 lbf/in$^2$ maximum. The UBC [Sec. ~~1603.5~~] $1612.3.2$ allows a one-third increase in allowable stresses for seismic forces.

$$A_s = \frac{T}{F_s} = \frac{T}{(0.5f_y)(1.33)}$$

For grade 60, $0.5f_y$ is 30,000 lbf/in$^2$; however, the maximum allowable stress is 24,000 lbf/in$^2$. Therefore,

$$A_s = \frac{39{,}600 \text{ lbf}}{\left(24{,}000\,\dfrac{\text{lbf}}{\text{in}^2}\right)(1.33)}$$

$$= 1.24 \text{ in}^2$$

**F.**

26. Answer C

*SI solution*

For the steel frames,

$$R = \frac{1}{\Delta} = \frac{1}{(25 \text{ mm})\left(\dfrac{1 \text{ in}}{25 \text{ mm}}\right)}$$

$$= 1$$

For the plywood shear walls,

$$R = \frac{1}{\Delta} = \frac{1}{(2.5 \text{ mm})\left(\dfrac{1 \text{ in}}{25 \text{ mm}}\right)}$$

$$= 10$$

$$R_{\text{total wall}} = R_{\text{steel frame}} + R_{\text{shear wall}} + R_{\text{steel frame}}$$
$$+ R_{\text{shear wall}} + R_{\text{steel frame}}$$
$$= 1 + 10 + 1 + 10 + 1$$
$$= 23$$

*Customary U.S. solution*

For the steel frames,

$$R = \frac{1}{\Delta} = \frac{1}{1.0}$$

$$= 1$$

For the plywood shear walls,

$$R = \frac{1}{\Delta} = \frac{1}{0.1}$$

$$= 10$$

$$R_{\text{total wall}} = R_{\text{steel frame}} + R_{\text{shear wall}} + R_{\text{steel frame}}$$
$$+ R_{\text{shear wall}} + R_{\text{steel frame}}$$
$$= 1 + 10 + 1 + 10 + 1$$
$$= 23$$

27. Answer D

*SI solution*

$$V = \left(\frac{R_{\text{shear wall}}}{R_{\text{wall}}}\right)(\text{lateral load})$$

$$= \left(\frac{10}{23}\right)(533\,784 \text{ N})$$

$$= 232\,080 \text{ N}$$

$$\approx 232\,000 \text{ N}$$

*Customary U.S. solution*

$$V = \left(\frac{R_{\text{shear wall}}}{R_{\text{wall}}}\right)(\text{lateral load})$$

$$= \left(\frac{10}{23}\right)(120{,}000 \text{ lbf})$$

$$= 52{,}174 \text{ lbf}$$

**28. Answer A**

*SI solution*

$$V = \left(\frac{R_{\text{steel frame}}}{R_{\text{wall}}}\right)(\text{lateral load})$$

$$= \left(\frac{1}{23}\right)(533\,784 \text{ N})$$

$$= 23\,208 \text{ N}$$

$$\approx 23\,210 \text{ N}$$

*Customary U.S. solution*

$$V = \left(\frac{R_{\text{steel frame}}}{R_{\text{wall}}}\right)(\text{lateral load})$$

$$= \left(\frac{1}{23}\right)(120{,}000 \text{ lbf})$$

$$= 5217 \text{ lbf}$$

**29. Answer A**

*SI solution*

| wall panels | each | $R$ |
|---|---|---|
| steel frame | 2 | 1 |
| plywood shear wall | 2 | 10 |
| concrete shear wall | 1 | 5 |

$$R_{\text{wall}} = (2)(1) + (2)(10) + (1)(5)$$

$$= 27$$

$$V = \left(\frac{R_{\text{concrete shear wall}}}{R_{\text{wall}}}\right)(\text{lateral load})$$

$$= \left(\frac{5}{27}\right)(533\,784 \text{ N})$$

$$= 98\,845 \text{ N}$$

*Customary U.S. solution*

| wall panels | each | $R$ |
|---|---|---|
| steel frame | 2 | 1 |
| plywood shear wall | 2 | 10 |
| concrete shear wall | 1 | 5 |

$$R_{\text{wall}} = (2)(1) + (2)(10) + (1)(5)$$

$$= 27$$

$$V = \left(\frac{R_{\text{concrete shear wall}}}{R_{\text{wall}}}\right)(\text{lateral load})$$

$$= \left(\frac{5}{27}\right)(120{,}000 \text{ lbf})$$

$$= 22{,}222 \text{ lbf}$$

**30. Answer B**

*SI solution*

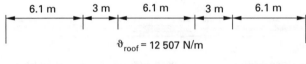

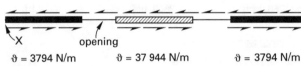

$$\vartheta_{\text{roof}} = \frac{V}{b}$$

$$= \frac{533\,784 \text{ N}}{42.5 \text{ m}}$$

$$= 12\,560 \text{ m}$$

From the solution to Prob. 27,

$$V_{\text{shear wall}} = 232\,000 \text{ N}$$

From the solution to Prob. 28,

$$V_{\text{steel frame}} = 23\,210 \text{ N}$$

$$\vartheta_{\text{steel frame}} = \frac{23\,210 \text{ N}}{6.1 \text{ m}}$$

$$= 3805 \text{ N/m}$$

$$\vartheta_{\text{shear wall}} = \frac{232\,000 \text{ N}}{6.1 \text{ m}}$$

$$= 38\,033 \text{ N/m}$$

Summing forces at X gives the drag strut force at that point.

$$\sum F_x = 0$$

$$D_X + \left(12\,560\ \frac{N}{m}\right)(24.3\ m)$$

$$-\left(3805\ \frac{N}{m}\right)(6.1\ m) - \left(38\,033\ \frac{N}{m}\right)(6.1\ m)$$

$$-\left(3805\ \frac{N}{m}\right)(6.1\ m) = 0$$

$$D_X = 26\,786\ N/m$$

$$\approx 26\,800\ N/m \quad [\text{tension}]$$

*Customary U.S. solution*

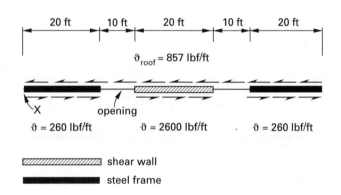

$\vartheta_{\text{roof}} = 857$ lbf/ft

$\vartheta = 260$ lbf/ft     $\vartheta = 2600$ lbf/ft     $\vartheta = 260$ lbf/ft

shear wall

steel frame

$$\vartheta_{\text{roof}} = \frac{V}{b}$$

$$= \frac{120,000\ \text{lbf}}{140\ \text{ft}}$$

$$= 857\ \text{lbf/ft}$$

From the solution to Prob. 27,

$$V_{\text{shear wall}} = 52,000\ \text{lbf}$$

From the solution to Prob. 28,

$$V_{\text{steel frame}} = 5200\ \text{lbf}$$

$$\vartheta_{\text{steel frame}} = \frac{5200\ \text{lbf}}{20\ \text{ft}}$$

$$= 260\ \text{lbf/ft}$$

$$\vartheta_{\text{shear wall}} = \frac{52,000\ \text{lbf}}{20\ \text{ft}}$$

$$= 2600\ \text{lbf/ft}$$

Summing forces at X gives the drag strut force at that point.

$$\sum F_x = 0$$

$$D_X + \left(857\ \frac{\text{lbf}}{\text{ft}}\right)(80\ \text{ft})$$

$$-\left(260\ \frac{\text{lbf}}{\text{ft}}\right)(20\ \text{ft}) - \left(2600\ \frac{\text{lbf}}{\text{ft}}\right)(20\ \text{ft})$$

$$-\left(260\ \frac{\text{lbf}}{\text{ft}}\right)(20\ \text{ft}) = 0$$

$$D_X = 6160\ \text{lbf} \quad [\text{tension}]$$

## G.

**31.** Answer D

*SI solution*

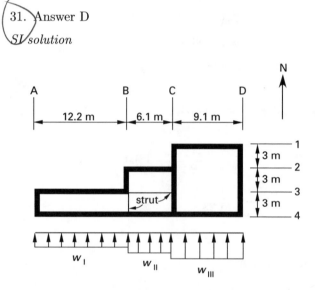

From UBC Table 16-I, $Z$ is 0.3. From UBC Table 16-N, $R$ is 5.5. From UBC Table 16-Q, the seismic coefficient $C_a$ is 0.24.

For determination of base shear based on the UBC Simplified Design Base Shear, use UBC Formula 30-11.

$$V = \left(\frac{3.0C_a}{R}\right)W$$

$$= \left(\frac{(3.0)(0.24)}{5.5}\right)W$$

$$= 0.13W$$

$w_\text{I}$ is the diaphragm loading for the section between lines A and B, $w_\text{II}$ is the diaphragm loading for the section between lines B and C, and $w_\text{III}$ is the diaphragm loading for the section between lines C and D.

Diaphragm loading $w_I$:

$$W_{\text{roof}} = (3 \text{ m})(12.2 \text{ m})\left(958 \, \frac{\text{N}}{\text{m}^2}\right)$$

$$= 35\,063 \text{ N}$$

$$W_{\text{walls}} = (2 \text{ walls})(12.2 \text{ m})\left(\frac{4.3 \text{ m}}{2}\right)\left(766 \, \frac{\text{N}}{\text{m}^2}\right)$$

$$= 40\,184 \text{ N}$$

$$W_{\text{total}} = 35\,063 \text{ N} + 40\,184 \text{ N}$$

$$= 75\,247 \text{ N}$$

$$V = (0.13)(75\,247 \text{ N})$$

$$= 9782 \text{ N}$$

The diaphragm loading is

$$w_I = \frac{9782 \text{ N}}{12.2 \text{ m}}$$

$$= 802 \text{ N/m}$$

Diaphragm loading $w_{II}$:

$$W_{\text{roof}} = (6.1 \text{ m})(6.1 \text{ m})\left(958 \, \frac{\text{N}}{\text{m}^2}\right)$$

$$= 35\,647 \text{ N}$$

$$W_{\text{walls}} = (2 \text{ walls})(6.1 \text{ m})\left(\frac{4.3 \text{ m}}{2}\right)\left(766 \, \frac{\text{N}}{\text{m}^2}\right)$$

$$= 20\,092 \text{ N}$$

$$W_{\text{total}} = (35\,647 \text{ N}) + (20\,092 \text{ N})$$

$$= 55\,739 \text{ N}$$

$$V = (0.13)(55\,739 \text{ N})$$

$$= 7246 \text{ N}$$

The diaphragm loading is

$$w_{II} = \frac{7246 \text{ N}}{6.1 \text{ m}}$$

$$= 1188 \text{ N/m}$$

Diaphragm loading $w_{III}$:

$$W_{\text{roof}} = (9.1 \text{ m})(9.1 \text{ m})\left(958 \, \frac{\text{N}}{\text{m}^2}\right)$$

$$= 79\,332 \text{ N}$$

$$W_{\text{walls}} = (2 \text{ walls})(9.1 \text{ m})\left(\frac{4.3 \text{ m}}{2}\right)\left(766 \, \frac{\text{N}}{\text{m}^2}\right)$$

$$= 29\,974 \text{ N}$$

$$W_{\text{total}} = 79\,332 \text{ N} + 29\,974 \text{ N}$$

$$= 109\,306 \text{ N}$$

$$V = (0.13)(109\,306 \text{ N})$$

$$= 14\,210 \text{ N}$$

The diaphragm loading is

$$w_{III} = \frac{14\,210 \text{ N}}{9.1 \text{ m}}$$

$$= 1562 \text{ N/m}$$

Roof shear force at line B is

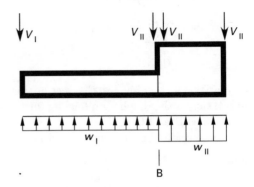

For a flexible diaphragm, the shear force is distributed to the parallel walls according to tributary areas. The tributary area to line B contributes half of the diaphragm force, $w_I$ and half of the diaphragm force, $w_{II}$.

$$V_{\text{shear}} = \frac{wL}{2}$$

$$V_B = \frac{w_I L_I}{2} + \frac{w_{II} L_{II}}{2}$$

$$= \frac{\left(802 \, \frac{\text{N}}{\text{m}}\right)(12.2 \text{ m})}{2} + \frac{\left(1188 \, \frac{\text{N}}{\text{m}}\right)(6.1 \text{ m})}{2}$$

$$= 8516 \text{ N}$$

*Customary U.S. solution*

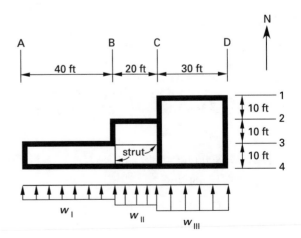

From UBC Table 16-I, $Z$ is 0.3. From UBC Table 16-N, $R$ is 5.5. From UBC Table 16-Q, the seismic coefficient $C_a$ is 0.24.

For determination of base shear based on the UBC Simplified Design Base Shear, use UBC Formula 30-11.

$$V = \left(\frac{3.0C_a}{R}\right) W$$
$$= \left(\frac{(3.0)(0.24)}{5.5}\right) W$$
$$= 0.13W$$

$w_{\mathrm{I}}$ is the diaphragm loading for the section between lines A and B, $w_{\mathrm{II}}$ is the diaphragm loading for the section between lines B and C, and $w_{\mathrm{III}}$ is the diaphragm loading for the section between lines C and D.

Diaphragm loading $w_{\mathrm{I}}$:

$$W_{\mathrm{roof}} = (10 \text{ ft})(40 \text{ ft})\left(20 \frac{\text{lbf}}{\text{ft}^2}\right)$$
$$= 8000 \text{ lbf}$$
$$W_{\mathrm{walls}} = (2 \text{ walls})(40 \text{ ft})\left(\frac{14 \text{ ft}}{2}\right)\left(16 \frac{\text{lbf}}{\text{ft}^2}\right)$$
$$= 8960 \text{ lbf}$$
$$W_{\mathrm{total}} = 8000 \text{ lbf} + 8960 \text{ lbf}$$
$$= 16{,}960 \text{ lbf}$$
$$V = (0.13)(16{,}960 \text{ lbf})$$
$$= 2205 \text{ lbf}$$

The diaphragm loading is

$$w_{\mathrm{I}} = \frac{2205 \text{ lbf}}{40 \text{ ft}}$$
$$= 55 \text{ lbf/ft}$$

Diaphragm loading $w_{\mathrm{II}}$:

$$W_{\mathrm{roof}} = (20 \text{ ft})(20 \text{ ft})\left(20 \frac{\text{lbf}}{\text{ft}^2}\right)$$
$$= 8000 \text{ lbf}$$
$$W_{\mathrm{walls}} = (2 \text{ walls})(20 \text{ ft})\left(\frac{14 \text{ ft}}{2}\right)\left(16 \frac{\text{lbf}}{\text{ft}^2}\right)$$
$$= 4480 \text{ lbf}$$
$$W_{\mathrm{total}} = 8000 \text{ lbf} + 4480 \text{ lbf}$$
$$= 12{,}480 \text{ lbf}$$
$$V = (0.13)(12{,}480 \text{ lbf})$$
$$= 1622 \text{ lbf}$$

The diaphragm loading is

$$w_{\mathrm{II}} = \frac{1622 \text{ lbf}}{20 \text{ ft}}$$
$$= 81 \text{ lbf/ft}$$

Diaphragm loading $w_{\mathrm{III}}$:

$$W_{\mathrm{roof}} = (30 \text{ ft})(30 \text{ ft})\left(20 \frac{\text{lbf}}{\text{ft}^2}\right)$$
$$= 18{,}000 \text{ lbf}$$
$$W_{\mathrm{walls}} = (2 \text{ walls})(30 \text{ ft})\left(\frac{14 \text{ ft}}{2}\right)\left(16 \frac{\text{lbf}}{\text{ft}^2}\right)$$
$$= 6720 \text{ lbf}$$
$$W_{\mathrm{total}} = 18{,}000 \text{ lbf} + 6720 \text{ lbf}$$
$$= 24{,}720 \text{ lbf}$$
$$V = (0.13)(24{,}720 \text{ lbf})$$
$$= 3214 \text{ lbf}$$

The diaphragm loading is

$$w_{\mathrm{III}} = \frac{3214 \text{ lbf}}{30 \text{ ft}}$$
$$= 107 \text{ lbf/ft}$$

Roof shear force at line B is

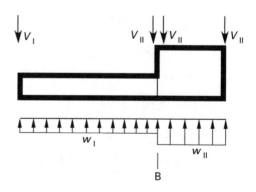

For a flexible diaphragm, the shear force is distributed to the parallel walls according to tributary areas. The tributary area to line B contributes half of the diaphragm force, $w_{\mathrm{I}}$ and half of the diaphragm force, $w_{\mathrm{II}}$.

$$V_{\text{shear}} = \frac{wL}{2}$$

$$V_{\text{B}} = \frac{w_{\text{I}}L_{\text{I}}}{2} + \frac{w_{\text{II}}L_{\text{II}}}{2}$$

$$= \frac{\left(55 \ \frac{\text{lbf}}{\text{ft}}\right)(40 \ \text{ft})}{2} + \frac{\left(81 \ \frac{\text{lbf}}{\text{ft}}\right)(20 \ \text{ft})}{2}$$

$$= 1910 \ \text{lbf}$$

32. Answer B

*SI solution*

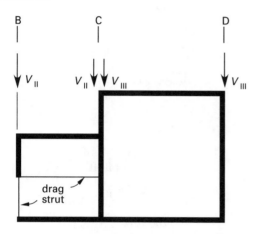

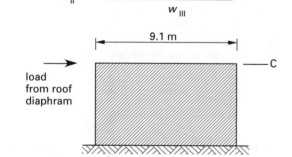

Shear force at line C:

$$V_{\text{roof}} = \frac{wL}{2}$$

$$V_{\text{roof}} = \frac{w_{\text{II}}L_{\text{II}}}{2} + \frac{w_{\text{III}}L_{\text{III}}}{2}$$

$$= \frac{\left(1188 \ \frac{\text{N}}{\text{m}}\right)(6.1 \ \text{m})}{2}$$

$$+ \frac{\left(1562 \ \frac{\text{N}}{\text{m}}\right)(9.1 \ \text{m})}{2}$$

$$V_{\text{roof}} = 10\,731 \ \text{N}$$

$$W_{\text{shear wall}} = (9.1 \ \text{m})\left(\frac{4.3 \ \text{m}}{2}\right)\left(766 \ \frac{\text{N}}{\text{m}^2}\right)$$

$$= 14\,987 \ \text{N}$$

$$V_{\text{shear wall}} = (0.13)(14\,987 \ \text{N})$$

$$= 1948 \ \text{N}$$

$$V = V_{\text{roof}} + V_{\text{shear wall}}$$

$$= 10\,731 \ \text{N} + 1948 \ \text{N}$$

$$= 12\,679 \ \text{N}$$

$$\vartheta_{\text{C}} = \frac{V}{b}$$

$$= \frac{12\,679 \ \text{N}}{9.1 \ \text{m}}$$

$$= 1393 \ \text{N/m}$$

*Customary U.S. solution*

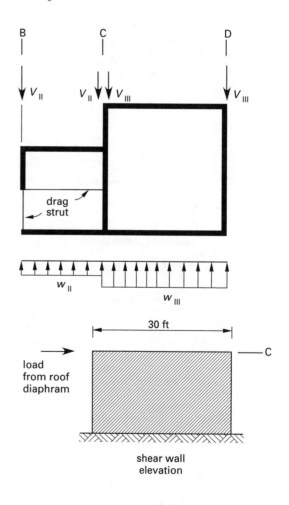

Shear force at line C:

$$V_{\text{roof}} = \frac{wL}{2}$$

$$V_{\text{roof}} = \frac{w_{\text{II}}L_{\text{II}}}{2} + \frac{w_{\text{III}}L_{\text{III}}}{2}$$

$$= \frac{\left(81 \, \dfrac{\text{lbf}}{\text{ft}}\right)(20 \text{ ft})}{2}$$

$$+ \frac{\left(107 \, \dfrac{\text{lbf}}{\text{ft}}\right)(30 \text{ ft})}{2}$$

$$V_{\text{roof}} = 2415 \text{ lbf}$$

$$W_{\text{shear wall}} = (30 \text{ ft})\left(\frac{14 \text{ ft}}{2}\right)\left(16 \, \frac{\text{lbf}}{\text{ft}^2}\right)$$

$$= 3360 \text{ lbf}$$

$$V_{\text{shear wall}} = (0.13)(3360 \text{ lbf})$$

$$= 437 \text{ lbf}$$

$$V = V_{\text{roof}} + V_{\text{shear wall}}$$

$$= 2415 \text{ lbf} + 437 \text{ lbf}$$

$$= 2852 \text{ lbf}$$

$$\vartheta_C = \frac{V}{b} = \frac{2852 \text{ lbf}}{30 \text{ ft}}$$

$$= 95 \text{ lbf/ft}$$

**33.** Answer C

*SI solution*

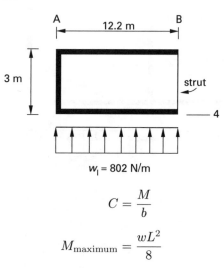

$$C = \frac{M}{b}$$

$$M_{\text{maximum}} = \frac{wL^2}{8}$$

Therefore,

$$C = \frac{wL^2}{8b}$$

$$= \frac{\left(802 \, \dfrac{\text{N}}{\text{m}}\right)(12.2 \text{ m})^2}{(8)(3 \text{ m})}$$

$$= 4974 \text{ N}$$

*Customary U.S. solution*

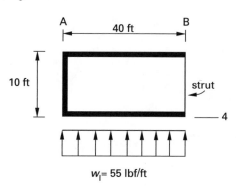

$$C = \frac{M}{b}$$

$$M_{\text{maximum}} = \frac{wL^2}{8}$$

Therefore,

$$C = \frac{wL^2}{8b}$$

$$= \frac{\left(55 \, \dfrac{\text{lbf}}{\text{ft}}\right)(40 \text{ ft})^2}{(8)(10 \text{ ft})}$$

$$= 1100 \text{ lbf}$$

**34.** Answer A

*SI solution*

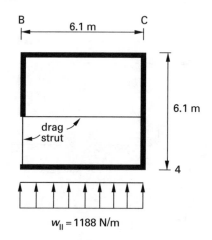

$$C = \frac{M}{b} = \frac{wL^2}{8b}$$

$$= \frac{\left(1188 \, \dfrac{\text{N}}{\text{m}}\right)(6.1 \text{ m})^2}{(8)(6.1 \text{ m})}$$

$$= 906 \text{ N}$$

*Customary U.S. solution*

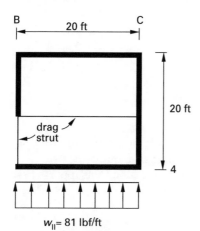

$w_{\text{II}}$= 81 lbf/ft

$$C = \frac{M}{b} = \frac{wL^2}{8b}$$

$$= \frac{\left(81 \dfrac{\text{lbf}}{\text{ft}}\right)(20 \text{ ft})^2}{(8)(20 \text{ ft})}$$

$$= 203 \text{ lbf}$$

**35. Answer B**

*SI solution*

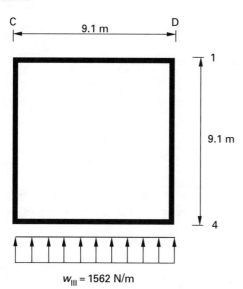

$w_{\text{III}}$ = 1562 N/m

$$C = \frac{M}{b} = \frac{wL^2}{8b}$$

$$= \frac{\left(1562 \dfrac{\text{N}}{\text{m}}\right)(9.1 \text{ m})^2}{(8)(9.1 \text{ m})}$$

$$= 1777 \text{ N}$$

*Customary U.S. solution*

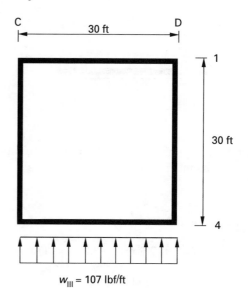

$w_{\text{III}}$ = 107 lbf/ft

$$C = \frac{M}{b} = \frac{wL^2}{8b}$$

$$= \frac{\left(107 \dfrac{\text{lbf}}{\text{ft}}\right)(30 \text{ ft})^2}{(8)(30 \text{ ft})}$$

$$= 401 \text{ lbf}$$

## H.

**36. Answer C**

*SI solution*

$$V = \frac{wL}{2}$$

$$= \frac{\left(2335 \dfrac{\text{N}}{\text{m}}\right)(4.6 \text{ m} + 7.6 \text{ m})}{2}$$

$$= 14\,244 \text{ N}$$

$$\vartheta_{\text{roof}} = \frac{V}{b_{\text{roof}}}$$

$$= \frac{14\,244 \text{ N}}{22.8 \text{ m}}$$

$$= 624.7 \text{ N/m}$$

$$\vartheta_{\text{wall}} = \frac{V}{b_{\text{walls}}}$$

$$= \frac{14\,244 \text{ N}}{7.6 \text{ m} + 7.6 \text{ m}}$$

$$= 937 \text{ N/m}$$

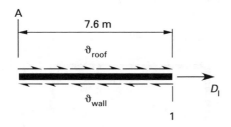

The force in the strut at point 1 can be determined by cutting the member and summing the forces.

$$\sum F_1 = 0$$

The strut force at 1 is

$$D_1 + \left(624.7 \ \frac{N}{m}\right)(7.6 \ m) - \left(937 \ \frac{N}{m}\right)(7.6 \ m) = 0$$

$$D_1 = 2374 \ N \quad \text{[tension]}$$

*Customary U.S. solution*

$$V = \frac{wL}{2} = \frac{\left(160 \ \frac{lbf}{ft}\right)(15 \ ft + 25 \ ft)}{2}$$

$$= 3200 \ lbf$$

$$\vartheta_{\text{roof}} = \frac{V}{b_{\text{roof}}} = \frac{3200 \ lbf}{75 \ ft}$$

$$= 42.7 \ lbf/ft$$

$$\vartheta_{\text{wall}} = \frac{V}{b_{\text{walls}}} = \frac{3200 \ lbf}{25 \ ft + 25 \ ft}$$

$$= 64 \ lbf/ft$$

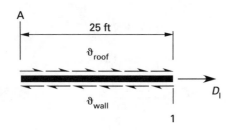

The force in the strut at point 1 can be determined by cutting the member and summing the forces.

$$\sum F_1 = 0$$

The strut force at 1 is

$$D_1 + \left(42.7 \ \frac{lbf}{ft}\right)(25 \ ft) - \left(64 \ \frac{lbf}{ft}\right)(25 \ ft) = 0$$

$$D_1 = 533 \ lbf$$

$$\approx 535 \ lbf \quad \text{[tension]}$$

**37. Answer A**

Along line Y, there is no unsupported horizontal diaphragm section that requires a strut. A strut drags the diaphragm shear along unsupported sections to the supporting shear wall.

**38. Answer C**

*SI solution*

From the solution to Prob. 36,

$$V_{\text{roof}} = 14\,244 \ N$$
$$\vartheta_{\text{roof}} = 624.7 \ N/m$$
$$\vartheta_{\text{wall}} = 937 \ N/m$$

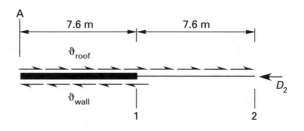

The force in the strut at point 2 can be determined by cutting the member and summing the forces.

$$\sum F_2 = 0$$

The strut force at 2 is

$$D_2 - \left(624.7 \ \frac{N}{m}\right)(15.2 \ m)$$

$$+ \left(937 \ \frac{N}{m}\right)(7.6 \ m) = 0$$

$$D_2 = 2374 \ N \quad \text{[compression]}$$

*Customary U.S. solution*

From the solution to Prob. 36,

$$V_{\text{roof}} = 3200 \text{ lbf}$$
$$\vartheta_{\text{roof}} = 42.7 \text{ lbf/ft}$$
$$\vartheta_{\text{wall}} = 64 \text{ lbf/ft}$$

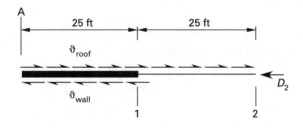

The force in the strut at point 2 can be determined by cutting the member and summing the forces.

$$\sum F_2 = 0$$

The strut force at 2 is

$$D_2 - \left(42.7 \frac{\text{lbf}}{\text{ft}}\right)(50 \text{ ft}) + \left(64 \frac{\text{lbf}}{\text{ft}}\right)(25 \text{ ft}) = 0$$
$$D_2 = 535 \text{ lbf} \quad \text{[compression]}$$

**39. Answer D**

*SI solution*

$$V = \frac{wL}{2}$$
$$= \frac{\left(5838 \frac{\text{N}}{\text{m}}\right)(7.6 \text{ m})}{2}$$
$$= 22\,184 \text{ N}$$
$$\vartheta_{\text{roof}} = \frac{V}{b_{\text{roof}}}$$
$$= \frac{22\,184 \text{ N}}{12.2 \text{ m}}$$
$$= 1818 \text{ N/m}$$

$$\vartheta_{\text{wall A}} = \frac{V}{b_{\text{walls}}}$$
$$= \frac{22\,184 \text{ N}}{7.6 \text{ m}}$$
$$= 2919 \text{ N/m}$$
$$\vartheta_{\text{wall B}} = \frac{V}{b_{\text{walls}}}$$
$$= \frac{22\,184 \text{ N}}{6.1 \text{ m} + 1.5 \text{ m}}$$
$$= 2919 \text{ N/m}$$

The strut force at point 3 along line B is

$$D_3 - \left(1818 \frac{\text{N}}{\text{m}}\right)(10.7 \text{ m}) + \left(2919 \frac{\text{N}}{\text{m}}\right)(6.1 \text{ m}) = 0$$
$$D_3 = 1647 \quad \text{[compression]}$$

The strut force at point 4 along line B is

$$D_4 + \left(1818 \frac{\text{N}}{\text{m}}\right)(6.1 \text{ m}) - \left(2919 \frac{\text{N}}{\text{m}}\right)(6.1 \text{ m}) = 0$$
$$D_4 = 6716 \text{ N} \quad \text{[tension]}$$

The strut force at point 5 is

$$D_5 = 0$$

The strut force at point 6 is

$$D_6 - \left(1818 \frac{\text{N}}{\text{m}}\right)(4.6 \text{ m}) = 0$$
$$D_6 = 8363 \text{ N} \quad \text{[compression]}$$

The magnitude of the strut force is greatest at point 6.

*Customary U.S. solution*

$$V = \frac{wL}{2}$$

$$= \frac{\left(400 \, \frac{\text{lbf}}{\text{ft}}\right)(25 \text{ ft})}{2}$$

$$= 5000 \text{ lbf}$$

$$\vartheta_{\text{roof}} = \frac{V}{b_{\text{roof}}}$$

$$= \frac{5000 \text{ lbf}}{40 \text{ ft}}$$

$$= 125 \text{ lbf/ft}$$

$$\vartheta_{\text{wall A}} = \frac{V}{b_{\text{walls}}}$$

$$= \frac{5000 \text{ lbf}}{25 \text{ ft}}$$

$$= 200 \text{ lbf/ft}$$

$$\vartheta_{\text{wall B}} = \frac{V}{b_{\text{walls}}}$$

$$= \frac{5000 \text{ lbf}}{20 \text{ ft} + 5 \text{ ft}}$$

$$= 200 \text{ lbf/ft}$$

The strut force at point 3 along line B is

$$D_3 - \left(125 \, \frac{\text{lbf}}{\text{ft}}\right)(35 \text{ ft}) + \left(200 \, \frac{\text{lbf}}{\text{ft}}\right)(20 \text{ ft}) = 0$$

$$D_3 = 375 \text{ lbf} \quad [\text{compression}]$$

The strut force at point 4 along line B is

$$D_4 + \left(125 \, \frac{\text{lbf}}{\text{ft}}\right)(20 \text{ ft}) - \left(200 \, \frac{\text{lbf}}{\text{ft}}\right)(20 \text{ ft}) = 0$$

$$D_4 = 1500 \text{ lbf} \quad [\text{tension}]$$

The strut force at point 5 is $D_5 = 0$.

The strut force at point 6 is

$$D_6 - \left(125 \, \frac{\text{lbf}}{\text{ft}}\right)(15 \text{ ft}) = 0$$

$$D_6 = 1875 \text{ lbf} \quad [\text{compression}]$$

The magnitude of the strut force is greatest at point 6.

**40. Answer C**

*SI solution*

From the solution to Prob. 36,

$$V_{\text{roof}} = 14\,244 \text{ N}$$

$$\vartheta_{\text{roof}} = \frac{V}{b_{\text{roof}}}$$

$$= \frac{14\,244 \text{ N}}{22.8 \text{ m}}$$

$$= 625 \text{ N/m}$$

The drag strut between points 1 and 2 is

$$\left(625 \, \frac{\text{N}}{\text{m}}\right)(7.6 \text{ m}) = 4750 \text{ N}$$

*Customary U.S. solution*

From the solution to Prob. 36,

$$V_{\text{roof}} = 3200 \text{ lbf}$$

$$\vartheta_{\text{roof}} = \frac{V}{b_{\text{roof}}}$$

$$= \frac{3200 \text{ lbf}}{75 \text{ ft}}$$

$$= 42.7 \text{ lbf/ft}$$

The drag strut between points 1 and 2 is

$$\left(42.7 \, \frac{\text{lbf}}{\text{ft}}\right)(25 \text{ ft}) = 1068 \text{ lbf}$$

**41.** Answer C

*SI solution*

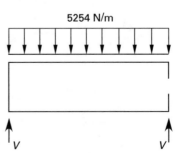

$$V = \frac{wL}{2}$$

$$= \frac{\left(5254 \; \frac{\text{N}}{\text{m}}\right)(15.2 \text{ m})}{2}$$

$$= 39\,930 \text{ N}$$

$$\vartheta = \frac{V}{b}$$

$$= \frac{39\,930 \text{ N}}{9.1 \text{ m}}$$

$$= 4388 \; \frac{\text{N}}{\text{m}}$$

The magnitude of the strut force over the 2.4 m opening is

$$\left(4388 \; \frac{\text{N}}{\text{m}}\right)(2.4 \text{ m}) = 10\,531 \text{ N} \quad \text{[tension or compression]}$$

*Customary U.S. solution*

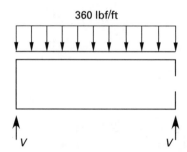

360 lbf/ft

$$V = \frac{wL}{2}$$

$$= \frac{\left(360 \; \frac{\text{lbf}}{\text{ft}}\right)(50 \text{ ft})}{2}$$

$$= 9000 \text{ lbf}$$

$$\vartheta = \frac{V}{b}$$

$$= \frac{9000 \text{ lbf}}{30 \text{ ft}}$$

$$= 300 \text{ lbf/ft}$$

The magnitude of the strut force over the 8 ft opening is

$$\left(300 \; \frac{\text{lbf}}{\text{ft}}\right)(8 \text{ ft}) = 2400 \text{ lbf} \quad \text{[tension or compression]}$$

**42.** Answer C

*SI solution*

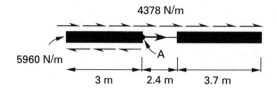

4378 N/m

5960 N/m

A

3 m    2.4 m    3.7 m

$$\vartheta_{\text{wall}} = \frac{V}{b} = \frac{39\,930 \text{ N}}{3.7 \text{ m} + 3.0 \text{ m}}$$

$$= 5960 \text{ N/m}$$

The strut force at point A is

$$\Sigma D_A = 0$$

$$D_A + \left(4378 \; \frac{\text{N}}{\text{m}}\right)(3.0 \text{ m}) - \left(5960 \; \frac{\text{N}}{\text{m}}\right)(3.0 \text{ m})$$

$$D_A = 4746 \quad \text{[tension]}$$

*Customary U.S. solution*

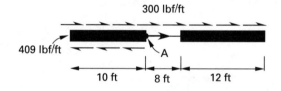

300 lbf/ft

409 lbf/ft

A

10 ft    8 ft    12 ft

$$\vartheta_{\text{wall}} = \frac{V}{b} = \frac{9000 \text{ lbf}}{12 \text{ ft} + 10 \text{ ft}}$$

$$= 409 \text{ lbf/ft}$$

The strut force at point A is

$$\sum D_A = 0$$

$$D_A + \left(300 \; \frac{\text{lbf}}{\text{ft}}\right)(10 \text{ ft}) - \left(409 \; \frac{\text{lbf}}{\text{ft}}\right)(10 \text{ ft}) = 0$$

$$D_A = 1090 \text{ lbf} \quad \text{[tension]}$$

**43.** Answer C

*SI solution*

The proportion of resistance is

$$\frac{3.0 \text{ m}}{3.0 \text{ m} + 3.7 \text{ m}} = 45\%$$

*Customary U.S. solution*

The proportion of resistance is

$$\frac{10 \text{ ft}}{10 \text{ ft} + 12 \text{ ft}} = 45\%$$

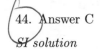

**44.** Answer C

*SI solution*

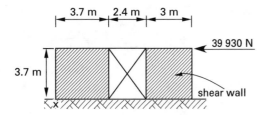

$$\vartheta_{\text{wall}} = \frac{V}{b} = \frac{39\,930 \text{ N}}{3.7 \text{ m} + 3.0 \text{ m}}$$
$$= 5960 \text{ N/m}$$

The overturning moment at X is

$$\left(5960 \ \frac{\text{N}}{\text{m}}\right)(3.7 \text{ m panel width})$$
$$\times (3.7 \text{ m panel height}) = 81\,592 \text{ N·m}$$
$$\approx 82 \text{ kN·m}$$

*Customary U.S. solution*

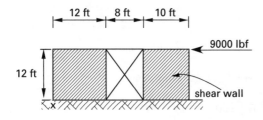

$$\vartheta_{\text{wall}} = \frac{V}{b} = \frac{9000 \text{ lbf}}{12 \text{ ft} + 10 \text{ ft}}$$
$$= 409 \text{ lbf/ft}$$

The overturning moment at X is

$$\left(409 \ \frac{\text{lbf}}{\text{ft}}\right)$$
$$\times (12 \text{ ft panel width})(12 \text{ ft panel height})$$
$$= 58,896 \text{ lbf-ft}$$

**45.** Answer C

*SI solution*

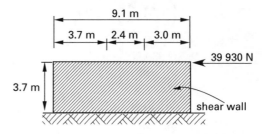

$$D_{\text{wall}} = (9.1 \text{ m})(3.7 \text{ m})\left(766 \ \frac{\text{N}}{\text{m}^2}\right)$$
$$= 25\,791 \text{ N}$$

To resist overturning, the UBC limits the dead load contribution to 90%.

The resisting moment is

$$(90\%)(25\,791 \text{ N})\left(\frac{9.1 \text{ ft}}{2}\right) = 105\,612.1 \text{ N·m}$$
$$\approx 106 \text{ kN·m}$$

*Customary U.S. solution*

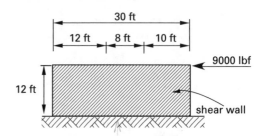

$$D_{\text{wall}} = (30 \text{ ft})(12 \text{ ft})\left(16 \ \frac{\text{lbf}}{\text{ft}^2}\right)$$
$$= 5760 \text{ lbf}$$

To resist overturning, the UBC limits the dead load contribution to 90%.

The resisting moment is

$$(90\%)(5760 \text{ lbf})\left(\frac{30 \text{ ft}}{2}\right) = 77\,760 \text{ ft-lbf}$$
$$\approx 78 \text{ ft-k}$$

# J.

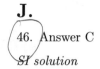

46. Answer C

*SI solution*

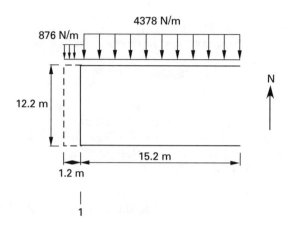

$$V = (1.2 \text{ m}) \left( 876 \ \frac{\text{N}}{\text{m}} \right) + \frac{(30.5 \text{ m}) \left( 4378 \ \frac{\text{N}}{\text{m}} \right)}{2}$$
$$= 67\,816 \text{ N}$$

*Customary U.S. solution*

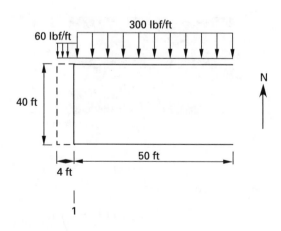

$$V = (4 \text{ ft}) \left( 60 \ \frac{\text{lbf}}{\text{ft}} \right) + \frac{(100 \text{ ft}) \left( 300 \ \frac{\text{lbf}}{\text{ft}} \right)}{2}$$
$$= 15,240 \text{ lbf}$$

47. Answer B

*SI solution*

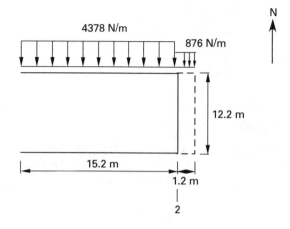

$$V = (1.2 \text{ m}) \left( 876 \ \frac{\text{N}}{\text{m}} \right) + \frac{(30.5 \text{ m}) \left( 4378 \ \frac{\text{N}}{\text{m}} \right)}{2}$$
$$= 67\,816 \text{ N}$$
$$\vartheta = \frac{V}{b} = \frac{67\,816 \text{ N}}{12.2 \text{ m}}$$
$$= 5559 \text{ N/m}$$
$$\approx 5560 \text{ N/m}$$

*Customary U.S. solution*

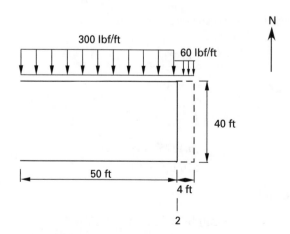

$$V = (4 \text{ ft}) \left( 60 \, \frac{\text{lbf}}{\text{ft}} \right) + \frac{(100 \text{ ft}) \left( 300 \, \frac{\text{lbf}}{\text{ft}} \right)}{2}$$

$$= 15{,}240 \text{ lbf}$$

$$\vartheta = \frac{V}{b} = \frac{15{,}240 \text{ lbf}}{40 \text{ ft}}$$

$$= 381 \text{ lbf/ft}$$

**48. Answer C**

*SI solution*

From the solution to Prob. 46, the roof shear force at line 1 is 67 816 N. The dead load of the parallel walls contributing to the wall shear stress is

$$W = (12.2 \text{ m}) \left( \frac{4.3 \text{ m}}{2} \right) \left( 766 \, \frac{\text{N}}{\text{m}^2} \right)$$

$$= 20\,092 \text{ N}$$

$$V = \left( \frac{3.0 C_a}{R} \right) W$$

$$= (0.1375)(20\,092 \text{ N})$$

$$= 2763 \text{ N}$$

The total shear is

$$67\,816 \text{ N} + 2763 \text{ N} = 70\,759$$

$$\vartheta = \frac{V}{b} = \frac{70\,579}{12.2 \text{ m}}$$

$$= 5785 \text{ N/m}$$

*Customary U.S. solution*

From the solution to Prob. 46, the roof shear force at line 1 is 15,240 lbf. The dead load of the parallel walls contributing to the wall shear stress is

$$W = (40 \text{ ft}) \left( \frac{14 \text{ ft}}{2} \right) \left( 16 \, \frac{\text{lbf}}{\text{ft}^2} \right)$$

$$= 4480 \text{ lbf}$$

$$V = \left( \frac{3.0 C_a}{R} \right) W$$

$$= (0.1375)(4480 \text{ lbf})$$

$$= 616 \text{ lbf}$$

The total shear is

$$15{,}240 \text{ lbf} + 616 \text{ lbf} = 15{,}856 \text{ lbf}$$

$$\vartheta = \frac{V}{b} = \frac{15{,}856 \text{ lbf}}{40 \text{ ft}}$$

$$= 396.4 \text{ lbf/ft}$$

**49. Answer C**

Due to symmetry, the diaphragm shear stress is the same as that determined in the solution to Prob. 47 for line 2.

$$\vartheta_{\text{roof}} = 381 \text{ lbf/ft} \quad (5560 \text{ N/m})$$

From UBC Table 23-II-H, the wood structural panel nail spacing at diaphragm boundaries necessary to achieve the allowable shear of at least 381 lbf/ft (5560 N/m) is 4 in (10 cm). At this spacing, the allowable shear is 425 lbf/ft (6202 N/m).

**50. Answer B**

From UBC Table 23-II-I-1, the nail spacing at wood structural panel edges should be 3 in (8 cm). For the given conditions and a nail spacing of 3 in (8 cm), the allowable shear is 410 lbf/ft (5983 N/m).

# K.

**51. Answer D**

*SI solution*

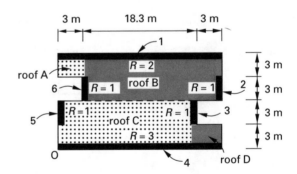

$$D_{\text{entire roof}} = \Big( \big[ (12 \text{ m})(24.3 \text{ m}) - (3 \text{ m})(3 \text{ m}) \big]$$

$$- (3 \text{ m})(3 \text{ m}) \Big) \left( 1197 \, \frac{\text{N}}{\text{m}^2} \right) \left( \frac{1 \text{ kN}}{1000 \text{ N}} \right)$$

$$= 327 \text{ kN}$$

For walls 1 and 4,

$$D = (24.3 \text{ m})(3.7 \text{ m}) \left( 2394 \, \frac{\text{N}}{\text{m}^2} \right) \left( \frac{1 \text{ kN}}{1000 \text{ N}} \right)$$

$$= 215 \text{ kN}$$

For walls 2, 3, 5, and 6,

$$D = (3 \text{ m})(3.7 \text{ m}) \left(2394 \, \frac{\text{N}}{\text{m}^2}\right) \left(\frac{1 \text{ kN}}{1000 \text{ N}}\right)$$
$$= 27 \text{ kN}$$

For roof A,

$$D = (3 \text{ m})(3 \text{ m}) \left(1197 \, \frac{\text{N}}{\text{m}^2}\right) \left(\frac{1 \text{ kN}}{1000 \text{ N}}\right)$$
$$= 11 \text{ kN}$$

For roof B,

$$D = (21.3 \text{ m})(6 \text{ m}) \left(1197 \, \frac{\text{N}}{\text{m}^2}\right) \left(\frac{1 \text{ kN}}{1000 \text{ N}}\right)$$
$$= 153 \text{ kN}$$

For roof C,

$$D = (21.3 \text{ m})(6 \text{ m}) \left(1197 \, \frac{\text{N}}{\text{m}^2}\right) \left(\frac{1 \text{ kN}}{1000 \text{ N}}\right)$$
$$= 153 \text{ kN}$$

For roof D,

$$D = (3 \text{ m})(3 \text{ m}) \left(1197 \, \frac{\text{N}}{\text{m}^2}\right) \left(\frac{1 \text{ kN}}{1000 \text{ N}}\right)$$
$$= 11 \text{ kN}$$

| | $D$ (kN) | $x$ (m) | $y$ (m) | $xD$ | $yD$ |
|---|---|---|---|---|---|
| roof A | 11 | 1.5 | 10.5 | 16.5 | 115.5 |
| B | 153 | 13.5 | 9.0 | 2065.5 | 1377.0 |
| C | 153 | 10.5 | 3.0 | 1606.5 | 459.0 |
| D | 11 | 22.8 | 1.5 | 250.8 | 16.5 |
| wall 1 | 215 | 12.0 | 12.0 | 2580.0 | 2580.0 |
| 2 | 27 | 24.3 | 7.5 | 656.1 | 202.5 |
| 3 | 27 | 21.3 | 4.5 | 575.1 | 121.5 |
| 4 | 215 | 12.0 | 0 | 2580.0 | 0 |
| 5 | 27 | 0 | 4.5 | 0 | 121.5 |
| 6 | 27 | 3.0 | 7.5 | 81 | 202.5 |
| total | $\sum D$ 866 kN | | | $\sum (D_i x_i)$ 10 411.5 kN·m | $\sum (D_i y_i)$ 5196.0 kN·m |

$$\overline{x} = \frac{\sum (D_i x_i)}{\sum D_i} = \frac{10\,411.5 \text{ kN·m}}{866 \text{ kN}}$$
$$= 12.0 \text{ m}$$

$$\overline{y} = \frac{\sum (D_i y_i)}{\sum D_i} = \frac{5196.0 \text{ kN·m}}{866 \text{ kN}}$$
$$= 6.0 \text{ m}$$

Note that for this problem, you could have guessed by inspection that the location of the center of mass would be at 12 m and 6 m in the $x$- and $y$-directions, due to symmetry.

*Customary U.S. solution*

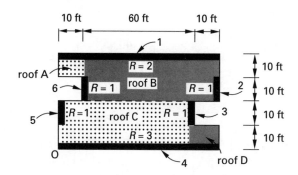

$$D_{\text{entire roof}} = \Big( [(40 \text{ ft})(80 \text{ ft}) - (10 \text{ ft})(10 \text{ ft})]$$
$$- (10 \text{ ft})(10 \text{ ft}) \Big) \left(25 \, \frac{\text{lbf}}{\text{ft}^2}\right)$$
$$= 75,000 \text{ lbf}$$
$$= 75 \text{ k}$$

For walls 1 and 4,

$$D = (80 \text{ ft})(12 \text{ ft}) \left(50 \, \frac{\text{lbf}}{\text{ft}^2}\right) \left(\frac{1 \text{ k}}{1000 \text{ lbf}}\right)$$
$$= 48 \text{ k}$$

For walls 2, 3, 5, and 6,

$$D = (10 \text{ ft})(12 \text{ ft}) \left(50 \, \frac{\text{lbf}}{\text{ft}^2}\right) \left(\frac{1 \text{ k}}{1000 \text{ lbf}}\right)$$
$$= 6 \text{ k}$$

For roof A,

$$D = (10 \text{ ft})(10 \text{ ft}) \left(25 \, \frac{\text{lbf}}{\text{ft}^2}\right) \left(\frac{1 \text{ k}}{1000 \text{ lbf}}\right)$$
$$= 2.5 \text{ k}$$

For roof B,

$$D = (70 \text{ ft})(20 \text{ ft}) \left(25 \, \frac{\text{lbf}}{\text{ft}^2}\right) \left(\frac{1 \text{ k}}{1000 \text{ lbf}}\right)$$
$$= 35 \text{ k}$$

For roof C,

$$D = (70 \text{ ft})(20 \text{ ft}) \left(25 \frac{\text{lbf}}{\text{ft}^2}\right) \left(\frac{1 \text{ k}}{1000 \text{ lbf}}\right)$$

$$= 35 \text{ k}$$

For roof D,

$$D = (10 \text{ ft})(10 \text{ ft}) \left(25 \frac{\text{lbf}}{\text{ft}^2}\right) \left(\frac{1 \text{ k}}{1000 \text{ lbf}}\right)$$

$$= 2.5 \text{ k}$$

|        | $D$ (k) | $x$ (ft) | $y$ (ft) | $xD$ | $yD$ |
|--------|---------|----------|----------|------|------|
| roof A | 2.5 | 5 | 35 | 12.5 | 87.5 |
| B | 35.0 | 45 | 30 | 1575.0 | 1050.0 |
| C | 35.0 | 35 | 10 | 1225.0 | 350.0 |
| D | 2.5 | 75 | 5 | 187.5 | 12.5 |
| wall 1 | 48.0 | 40 | 40 | 1920.0 | 1920.0 |
| 2 | 6.0 | 80 | 25 | 480.0 | 150.0 |
| 3 | 6.0 | 70 | 15 | 420.0 | 90.0 |
| 4 | 48.0 | 40 | 0 | 1920.0 | 0.0 |
| 5 | 6.0 | 0 | 15 | 0.0 | 90.0 |
| 6 | 6.0 | 10 | 25 | 60.0 | 150.0 |
| total | $\sum D$ 195 k | | | $\sum (D_i x_i)$ 7800 ft-k | $\sum (D_i y_i)$ 3900 ft-k |

$$\overline{x} = \frac{\sum (D_i x_i)}{\sum D_i} = \frac{7800 \text{ ft-k}}{195 \text{ k}}$$
$$= 40 \text{ ft}$$
$$\overline{y} = \frac{\sum (D_i y_i)}{\sum D_i} = \frac{3900 \text{ ft-k}}{195 \text{ k}}$$
$$= 20 \text{ ft}$$

Note that, for this problem, you could have guessed by inspection that the location of the center of mass would be at 40 ft and 20 ft in the $x$- and $y$-directions, due to symmetry.

## 52. Answer B

*SI solution*

| wall | $R$ | $x$ (m) | $y$ (m) |
|------|-----|---------|---------|
| 1 | 2 | 12.2 | 12.0 |
| 2 | 1 | 24.3 | 7.5 |
| 3 | 1 | 21.3 | 4.5 |
| 4 | 3 | 12.2 | 0 |
| 5 | 1 | 0 | 4.5 |
| 6 | 1 | 3.0 | 7.5 |

Assume a north-south loading direction and omit the weak walls (walls 1 and 4, in this case).

$$\overline{x}_R = \frac{\sum R_i x_i}{\sum R_i}$$
$$= \frac{(1)(0.0) + (1)(3.0) + (1)(21.3) + (1)(24.3)}{1 + 1 + 1 + 1}$$
$$= 12.0 \text{ m}$$

Assume an east-west loading direction and omit the weak walls (walls 2, 3, 5, and 6, in this case).

$$\overline{y}_R = \frac{\sum R_i y_i}{\sum R_i}$$
$$= \frac{(2)(12.0) + (3)(0)}{2 + 3}$$
$$= 4.8 \text{ m}$$

*Customary U.S. solution*

| wall | $R$ | $x$ (ft) | $y$ (ft) |
|------|-----|----------|----------|
| 1 | 2 | 40 | 40 |
| 2 | 1 | 80 | 25 |
| 3 | 1 | 70 | 15 |
| 4 | 3 | 40 | 0 |
| 5 | 1 | 0 | 15 |
| 6 | 1 | 10 | 25 |

Assume a north-south loading direction and omit the weak walls (walls 1 and 4, in this case).

$$\overline{x}_R = \frac{\sum R_i x_i}{\sum R_i}$$
$$= \frac{(1)(0) + (1)(10) + (1)(70) + (1)(80)}{1 + 1 + 1 + 1}$$
$$= 40 \text{ ft}$$

Assume an east-west loading direction and omit the weak walls (walls 2, 3, 5, and 6, in this case).

$$\overline{y}_R = \frac{\sum R_i y_i}{\sum R_i} = \frac{(2)(40) + (3)(0)}{2 + 3}$$
$$= 16 \text{ ft}$$

## 53. Answer B

*SI solution*

$$e_x = \overline{x}_C - \overline{x}_R = 12.0 - 12.0$$
$$= 0 \text{ m}$$
$$e_y = \overline{y}_C - \overline{y}_R = 6.0 - 4.8$$
$$= 1.2 \text{ m}$$
$$= 120 \text{ cm}$$

*Customary U.S. solution*

$$e_x = \overline{x}_C - \overline{x}_R = 40 \text{ ft} - 40 \text{ ft}$$
$$= 0 \text{ ft}$$

$$e_y = \overline{y}_C - \overline{y}_R = 20 \text{ ft} - 16 \text{ ft}$$
$$= 4 \text{ ft}$$

## 54. Answer B

*SI solution*

Per the UBC [Secs. 1630.6 and 1630.7], the accidental torsional moment is

$$e_a = L(5\%)$$
$$= (24.3 \text{ m})(5\%)$$
$$= 1.2 \text{ m}$$
$$T_x = Ve = V(e_x + e_a)$$
$$= (890 \text{ kN})(0 \text{ m} + 1.2 \text{ m})$$
$$= 1068 \text{ kN·m}$$

*Customary U.S. solution*

Per the UBC [Secs. 1630.6 and 1630.7], the accidental torsional moment is

$$e_a = L(5\%) = (80 \text{ ft})(5\%)$$
$$= 4 \text{ ft}$$

$$T_x = Ve = V(e_x + e_a)$$
$$= (200 \text{ k})(0 \text{ ft} + 4 \text{ ft})$$
$$= 800 \text{ ft-k}$$

## 55. Answer C

*SI solution*

Per the UBC [Secs. 1630.6 and 1630.7], the accidental torsional moment is

$$e_a = L(5\%)$$
$$= (12 \text{ m})(5\%)$$
$$= 0.6 \text{ m}$$
$$T_y = Ve = V(e_y + e_a)$$
$$= (890 \text{ kN})(1.2 \text{ m} + 0.6 \text{ m})$$
$$= 1602 \text{ kN·m}$$

*Customary U.S. solution*

Per the UBC [Secs. 1630.6 and 1630.7], the accidental torsional moment is

$$e_a = L(5\%) = (40 \text{ ft})(5\%)$$
$$= 2 \text{ ft}$$

$$T_y = Ve = V(e_y + e_a)$$
$$= (200 \text{ k})(4 \text{ ft} + 2 \text{ ft})$$
$$= 1200 \text{ ft-k}$$

# APPENDIX A
## Useful Conversion Factors

| Multiply | By | To Obtain | Multiply | By | To Obtain |
|----------|-----|-----------|----------|-----|-----------|
| cm | 0.3937 | in | kg | 2.2046 | lbm |
| ft | 12 | in | kip | 1000 | lbf |
| ft | 0.3048 | m | kPa | 0.1450 | lbf/in$^2$ |
| ft$^2$ | 144 | in$^2$ | lbf | 4.4482 | N |
| ft$^3$ | 7.4810 | gal | lbf/ft | 14.5938 | N/m |
| ft-kips | 1.3558 | kN·m | lbf/ft$^2$ | 47.8803 | N/m$^2$ |
| ft-lbf | 1.3558 | N·m | lbf/ft$^2$ | 144 | lbf/in$^2$ |
| ft/sec | 0.3048 | m/s | lbf/in$^2$ | 6894.8 | Pa |
| gal | 0.1337 | ft$^3$ | lbf/in$^2$ | 6.8948 | kPa |
| gal | 0.0038 | m$^3$ | lbm | 0.4536 | kg |
| gravities | 32.2 | ft/sec$^2$ | m | 3.2808 | ft |
| gravities | 9.81 | m/s$^2$ | m$^2$ | 10.7639 | ft$^2$ |
| in | 2.54 | cm | mm | 0.0394 | in |
| in | 0.0833 | ft | N·m | 0.7376 | ft-lbf |
| in | 25.4 | mm | m/s$^2$ | 0.1019 | gravities |
| in$^2$ | 6.4516 | cm$^2$ | N | 0.2248 | lbf |
| in$^3$ | 16.3871 | cm$^3$ | Pa | $1.4505 \times 10^{-4}$ | lbf/in$^2$ |

# APPENDIX B
## References and Suggested Reading

Following is a combination list of references recommended by the California Board for Professional Engineers and Land Surveyors and those used by the author. The references identified with an asterisk, in addition to the *Uniform Building Code*, should be considered as suggested material by the Board for the seismic principles section of the Special Civil/Seismic Principle Examination.

## Required Reference:

International Conference of Building Officials. *Uniform Building Code*. Whittier, CA: International Conference of Building Officials, 1997. www.icbo.org
Telephone: (310) 699-0541

## Supplemental References:

*Ambrose, James S. and Dimistry Vergun. *Simplified Building Design for Wind and Earthquake Forces*, 3rd ed. New York: John Wiley & Sons, 1997.

Baradar, Majid. *Seismic Principles Practice Exams for the California Special Civil Engineer Examination*, 2nd ed. Belmont, CA: Professional Publications, Inc., 1999. www.ppi2pass.com

*Brandow, Gregg and Gary Hart. *Design of Concrete Masonry Structures*. California and Nevada Concrete Masonry Association.

*Breyer, Donald E. *Design of Wood Structures*, 3rd ed. New York: McGraw Hill Inc., 1993.

*California Board for Professional Engineers and Land Surveyors. *Professional Engineers Act*, Business and Professions Code Sections 6700-6799. Sacramento: California Board for Professional Engineers and Land Surveyors. www.dca.ca.gov/pels/

Department of the Army, the Navy, and the Air Force. *Technical Manual Seismic Design for Buildings*. Washington, D.C.

Faherty, Keith F. and Thomas G. Williamson. *Wood Engineering and Construction Handbook*, 3rd ed. New York: McGraw Hill Inc., 1998.

Lindeburg, Michael R. *Seismic Design of Building Structures*, 7th ed. Belmont, CA: Professional Publications Inc., 1996. www.ppi2pass.com

Naeim, Farzad. *The Seismic Design Handbook*. New York: Van Nostrand Reinhold, 1989.

Pauley, T. and M. J. N. Priestley. *Seismic Design of Reinforced Concrete and Masonry Buildings*. New York: John Wiley & Sons, Inc. 1992.

Seismic Safety Commission. *The Homeowners Guide to Earthquake Safety* and *The Commercial Property Owners Guide to Earthquake Safety*. CA: Seismic Safety Commission, 1992.

*Structural Engineers Association of California. *Recommended Lateral Force Requirements and Commentary*, 6th ed. ("Bluebook"). San Francisco: Seismology Committee Structural Engineers Association of California, 1996. www.seaoc.org